RETHINKING ENVIRONMENTAL MANAGEMENT IN THE PACIFIC RIM

Rethinking Environmental Management in the Pacific Rim

Exploring local participation in Bangkok, Thailand

AMRITA DANIERE
University of Toronto, Ontario, Canada
LOIS M. TAKAHASHI
University of California at Los Angeles, USA

Routledge
Taylor & Francis Group

LONDON AND NEW YORK

First published 2002 by Ashgate Publishing

Reissued 2018 by Routledge
2 Park Square, Milton Park, Abingdon, Oxon OX14 4RN
711 Third Avenue, New York, NY 10017, USA

Routledge is an imprint of the Taylor & Francis Group, an informa business

Notice:
Product or corporate names may be trademarks or registered trademarks, and are used only for identification and explanation without intent to infringe.

Publisher's Note
The publisher has gone to great lengths to ensure the quality of this reprint but points out that some imperfections in the original copies may be apparent.

Disclaimer
The publisher has made every effort to trace copyright holders and welcomes correspondence from those they have been unable to contact.

Typeset by Martingraphix, Cape Town, South Africa

A Library of Congress record exists under LC control number: 2001099671

ISBN 13: 978-1-138-73344-2 (hbk)
ISBN 13: 978-1-138-73340-4 (pbk)
ISBN 13: 978-1-315-18711-2 (ebk)

Contents

List of Tables

List of Figures

Acknowledgements

The authors are grateful to Professors Orathai Ard-Am and Anuchat Poungsomlee of the Institute for Population and Social Research, Mahidol University, Nakorn Pathom, Thailand for their generous advice and assistance with collecting information regarding environmental issues in Thailand. In addition, we wish to thank many individuals at the Thailand Development Research Institute Foundation in Bangkok without whom our initial knowledge about Thailand, and Bangkok in particular, could not have been gained. These include Chalongphop Sussangkarn, Sauwalak Kittiprapas, Mingsarn Kaosa-ard, Somthawin Patanavich, Banasopit Mekvichai and Suganya Hutaserani. Amrita Daniere sincerely appreciates their outstanding and entertaining efforts to educate a naive researcher from the United States.

In addition, we thank the following graduate students for their untiring efforts to help this book reach completion. They include Nadia Abu-Zahra, Helen Collins, Molly Davidson-Welling, John Iveson, Yael Levitte and Kate Swanson, all at the University of Toronto and Stacy Harwood who, at the time, was a student at the University of California, Irvine.

We also thank Theo Panayotou and John Montgomery at Harvard University, Sora Park Tanjasiri at the University of California Irvine, Randall Crane at UCLA and Charly Mehl in Bangkok, Thailand for their wisdom and support throughout the course of this project.

A number of institutions have provided financial support for our research including the Pacific Basin Research Center of Soka University, the University of California's Pacific Rim Research Program, the Social Science and Humanities Research Council of Canada and the Department of Geography at Middlebury College which provided a productive and beautiful sabbatical home for one of the authors. The authors also thank the Department of Geography at the University of Toronto, the Department of Urban Planning at UCLA, and the Department of Urban and Regional Planning at UC Irvine for their support at various points of the project.

Finally, we also thank the following publishers for permission to use slightly modified versions of previously published material: Edward Elgar Publications for material published in Social Capital and Well-being in Developing Economies, edited by Jonathan Isham, Sunder Ramaswamy and Thomas Kelly (2002); University of Chicago Press for material published in Economic Development and Cultural Change; Pion for material published in Environment and Planning C; Elsevier Science for material published in Habitat International and Kluwer Academic Publishers for Policy Sciences.

Chapter 1

Economic Growth and Urban Environmental Degradation in the Pacific Rim

At the beginning of the twenty-first century, it is impossible to find an urban enclave or metropolitan region that does not have grave concerns about the quality of its surrounding environment. Many believe that the degradation of urban environments is most devastating and overwhelming in rapidly growing urban areas of developing countries. In these urban spaces, development, linked to the economic and political reality of globalization, has been accompanied by unprecedented environmental destruction and public health concerns. Since the late 1980s, the Pacific Rim has experienced one of the world's fastest and most comprehensive economic and physical transitions. Not surprisingly, many cities along the Pacific Rim currently face environmental crises created by rapid and uncontrolled economic and physical growth.

While many rural areas within the Pacific Rim continue to be exploited for minerals and timber or are cultivated for agricultural exports, a great deal of the manufacturing and processing growth within the region is centred in existing cities. 'Uneven development' broadly characterizes the nature of growth and change in countries within the Southeast Asia (Figure 1.1); however, the damage to the natural environment apparent in cities is one of the most troubling aspect of the area's development pattern.[1] While deforestation and habitat deterioration are widely acknowledged in terms of their significance, environmental degradation in cities also has vital global, regional and local implications. Air and water pollution in Southeast Asian cities, for example, have obvious impacts for their metropolitan regions as well as for the entire globe because of their effects on complex and interconnected physical and biological systems. In addition to these, environmental issues such as solid waste disposal, pest and vermin management, and noise pollution have significant impacts on the health and quality of life of Southeast Asian urban residents at the local, regional and national levels.

---

[1] Dixon and Drakakis-Smith (1997) elaborate on both the form and meaning of Southeast Asia's uneven development in their volume.

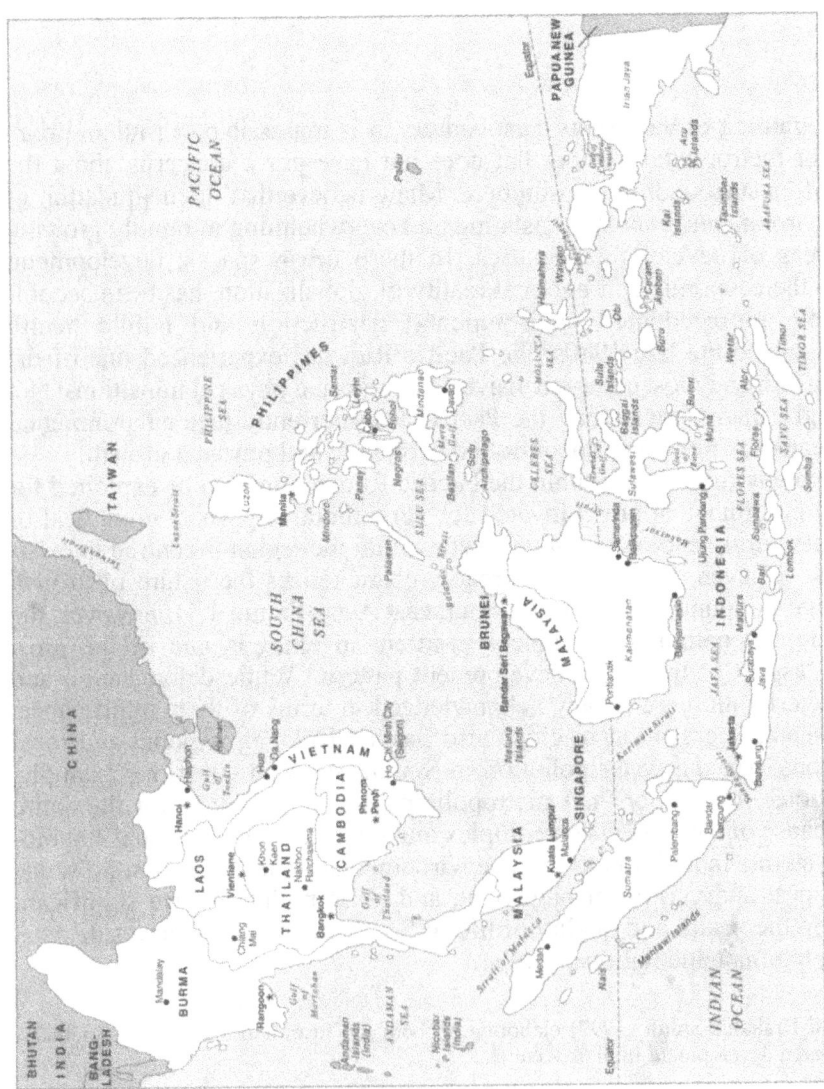

Figure 1.1 Map of Southeast Asia

Source: www.southeastasianews.net/map.html

It is our belief that the uneven development of Southeast Asian countries and other aspects of the region's patterns of growth and change are not unique. Rather, many of the features highlighted in discussions about Southeast Asian cities have much in common with the developing world as a whole. We focus most of the analysis in this book on a particular city in a specific Southeast Asian country to illuminate the important similarities between Bangkok and Thailand and much of the developing, and some of the developed, world. Perhaps no place on earth better epitomizes the phenomenon of rapid economic and environmental transition with few resources devoted to protection and planning, than the metropolitan area of Bangkok.

To provide a context for and comparison to the Bangkok case, however, the remainder of this chapter summarizes the recent experience of economic growth accompanied by urban environmental degradation in many of Southeast Asia's most important cities. We begin with a portrait of current economic conditions followed by a discussion of different types of environmental problems including water, sanitation, solid waste and air quality and the types of problems that governments face in trying to contain contamination. The following chapter also takes a broader view of environmental management conditions and issues within Southeast Asia. The discussion in both Chapters 1 and 2 is more general than many of the remaining chapters because they illustrate the relevance of the specific case of Thailand to the broader landscape of Southeast Asia and, more importantly, to the rest of the developing world.

Economic Conditions

The countries of Southeast Asia, which include Cambodia, Indonesia, Laos, Malaysia, Myanmar, the Philippines, Singapore, Thailand and Vietnam, comprise a wide range of development levels and types. This diversity of development patterns can be illustrated through the analysis of conventional economic indicators such as rates of economic growth or the composition of gross domestic product (GDP). Many of the nations in the region experienced relatively high growth rates (with the primary exceptions of the Philippines and Myanmar) since the early 1980s until the recent economic crisis in 1997. While the historic rates of growth have differed quite noticeably across and within countries, perhaps what stands out the most is that many Southeast Asian nations have experienced growth despite a myriad of economic, political and cultural differences (Table 1.1).

While the area is far from uniform in terms of development, there are quite a few features in common. Most striking since the early 1970s are the common themes of rapid development of industry, massive inflows of foreign direct investment, the emergence of authoritative regimes and the development of complex economic bureaucracies. In particular, and until the financial crisis that swept the region in 1997 and 1998, the success in Southeast Asia of policies emphasizing manufacturing has been exceptional. If we examine the region overall, annual average growth in gross domestic product (GDP) between 1980 and 2000 reveals how dynamic various economies have been. The strongest growth was recorded in Thailand (8.2 per cent per year) followed by Singapore (6.9 per cent), Malaysia (6.2 per cent) and Indonesia (5.8 per cent). Most economists and policy-makers familiar with the region agree that the Philippines economy performed the worst of all the ASEAN[2] members given the expectations most experts had of this country's capacity to develop once the Marcos regime was overthrown. The Philippines economy grew by only 1.4 per cent during the past two decades although more recently the economy has shown some signs of improvement.

Table 1.1 Economic Growth: Annual Average Percentage Growth of GDP

Country	1960 – 69[a]	1970 – 79[b]	1980 – 89[b]	1990 – 96[b]	1997 – 98[b]
Cambodia	n/a	n/a	n/a	6.23	0.99
Indonesia	3.90	7.66	5.32	8.02	-4.47
Laos	n/a	3.72	6.03	6.52	5.77
Malaysia	6.50	7.76	5.81	8.80	0.50
Myanmar	2.60	3.98	1.94	5.54	7.00
Philippines	5.1	6.16	2.00	2.83	2.31
Singapore	8.8	9.56	7.41	8.60	4.68
Thailand	8.2	7.00	7.35	8.50	-5.33
Vietnam	n/a	5.07	5.04	7.91	5.83

[a] *Source*: World Bank World Tables 1980, Economic Program Department, Washington, DC 1981.
[b] *Source*: International Monetary Fund, World Economic Outlook, World Economic Studies Division, Research Department, September 1999. Figures for recent years are IMF Staff Estimates.

Foreign direct investment is often credited with playing a major role in the development of almost all the countries in the region from the showpiece city-state of Singapore to the rapidly transforming Vietnamese economy. Many

[2] The Association of Southeast Asian Nations (ASEAN) was formed in 1967 to promote regional peace and security. The original members included Singapore, Malaysia, Indonesia, the Philippines and Thailand but more recently membership has expanded to include Brunei Darussalam, Vietnam, Burma, Cambodia and Laos.

believe that much of the economic growth experienced in the NIEs (newly industrializing economies) of Thailand, the Philippines, Malaysia and Indonesia was fuelled by vast sums of foreign direct investments such as in General Motors' 1996 decision to open a $750 million assembly plan on the eastern seaboard south of Bangkok, or the fact that in the early 1990s Malaysia was home to more than 250 Japanese companies. Initial interpretations of the late 1990s financial crisis, in fact, blamed footloose foreign investment for the speed and severity of the crisis in many Southeast Asian economies. Many argued that weaknesses in several key export industries, especially semiconductors, combined with intense competition from other strong exporting economies, particularly China, led to a devaluation of Southeast Asian curren-. cies. As Southeast Asian currencies fell in value relative to the US dollar, one positive effect was that the region's exports became more competitively priced on world markets. Unfortunately, the devaluations also motivated a substantial number of foreign investors to divest from the region. According to many analysts, this outflow worsened and deepened the region's financial crisis and led to projections of even slower regional economic growth.[3]

Interestingly, it appears that the absolute amount of foreign direct investment in a particular country is only somewhat associated with the magnitude of the effect of the so-called 'Asian flu' on the growth rate. While it is true that the countries most negatively impacted by the economic crisis, that is, Malaysia, Thailand and Indonesia, had large levels of investment both from foreign and domestic sources, a more accurate interpretation is that foreign direct investment (FDI) had a limited role in the impact of the 1997 and 1998 Asian economic downturns. Smaller and less developed economies such as Laos, where FDI made up almost 18 per cent of total investment, were less affected by the economic crises of 1997 and 1998 than larger, more export-oriented economies such as Thailand, where FDI accounted for about 7 per cent of all investments (Table 1.2). Apparently, the impacts of the economic downturn were more widely felt in larger Asian economies, particularly in those economies also experiencing structural political problems such as Indonesia. Smaller economies are often based more centrally on resource extraction, and linked more tenuously to the global economy, while larger economies in the region focused more on manufacturing and services. This in turn meant that the NIEs, in particular, were more vulnerable to weakening demand in specific manufacturing sectors and to poor decision-making on the part of over-optimistic central banks. We will return in the next chapter to the impacts of these economic downturns on the nature and role of environmental management in the region.

---

[3] See, for example, McLeod and Garnaut (1998).

Table 1.2 Foreign Direct Investment and GDP Growth Rate, 1998

Country	Total FDI (1990 – 97) (millions of US dollars)[a]	FDI (1997) as percentage of total investment that year in the country's economy[a]	GDP growth rate 1998[b] %
Cambodia	804	41.4	1.0
Indonesia	23684	7.0	−13.7
Laos	392	17.9	4.0
Malaysia	35178	12.1	−6.7
Myanmar	1106	n/a	7.0
Philippines	8348	6.0	−1.0
Singapore	49181	24.0	0.0
Thailand	17890	7.0	−9.4
Vietnam	6595	25.0	17.6

[a] *Source*: *World Development Indicators*, 1999 World Bank (CD-ROM)
[b] *Source*: International Monetary Fund, *World Economic Outlook*, World Economic Studies Division, Research Department, September 1999. Figures for recent years are IMF Staff Estimates.

Southeast Asian countries also diverge significantly as well in terms of GDP structure and, of course, living conditions. As indicated in Table 1.3, Myanmar is almost completely dependent on resource extraction (that is, logging and agriculture) for income generation while, at the other extreme, Singapore relies almost entirely on the service sector for employment and income generation. The same contrast holds true for measures of human welfare such as the GDP per capita or the infant mortality rate (Table 1.4). Singapore represents the positive end of the spectrum within the region with GDP per capita rates among the highest in the world and extremely low infant mortality rates (at 5 per 1000 live births which is the same as Switzerland and equal to the world's lowest rate). In contrast, Myanmar and Cambodia have GDP per capita rates of about US$200 per person (substantially lower than the levels for India and Bangladesh) and infant mortality rates of around 100 per 1000 live births (higher than rates reported for Bangladesh).

Table 1.3 Value Added as a Percentage of GDP, 1998

Country	Primary	Industry	Manufacturing	Services
Cambodia	51	15	6	34
Indonesia	16	43	26	41
Laos	52	21	16	27
Malaysia	12	48	34	40
Myanmar	59	10	7	31
Philippines	17	32	22	52
Singapore	0	35	24	65
Thailand	11	40	29	49
Vietnam	26	31	n.a.	43

Source: World Bank (1999). World Development Report 1999 – 2000. New York: Oxford University Press.

Table 1.4 Living Conditions, 1997

Country	GDP per capita US$	Average annual population growth rate (%) 1990–1998	Crude death rate per 1000	Infant mortality rate per 1000	Population over 65 (%)
Cambodia	293.89	3.1	12	103	3.0
Indonesia	1071.91	1.9	8	47	4.4
Laos	199.17	3.0	14	98	3.6
Malaysia	4545.37	2.8	5	11	3.9
Myanmar	200[a]	1.3	10	79	4.5
Philippines	1135.66	2.6	6	35	3.5
Singapore	30261.93	2.2	4	4	6.5
Thailand	2453.74	1.4	7	33	5.2
Vietnam	350.41	2.3	7	29	4.8

[a] The value published for Myanmar is an estimate based on what is an average value for a very low income country according to the World Bank. This value is very different from that reported by the IMF (that is, US$3474.07) and based on data submitted by Myanmar's government.

Sources: GDP per capita: IMF World Economic Outlook Database, September 1999. Available at www.imf.org/external/pubs/ft/weo/1999/02/data/index.htm

Others: World Bank World Development Indicators 1999 and World Development Report 1999 – 2000. Available at www.worldbank.org/data/wdi/home.html and www.worldbank.org/wdr.2000/fullreport.html

Political Structure and Economic Growth

Historically, the patterns of divergent development have their roots in the period following independence starting in the 1960s (Dixon, 1991, p. 21). Until the mid- to late 1980s, the growth rates of countries in the region diverged significantly among Singapore, the four core members of the Association for Southeast Asian States (ASEAN) that is, Indonesia, Malaysia, Thailand and the Philippines, and the communist and socialist nations. In the latter half of the 1980s and through the 1990s, however, changes in both economic structure and investment policies accelerated growth in Thailand, Malaysia and Indonesia that led to a decrease in the differences between these nations and Singapore. Concurrently, a dramatic shift in political and economic openness in Vietnam and Laos also reduced differences in the growth rates between 'Indochina' and the other countries in Southeast Asia.

As documented extensively elsewhere (Dixon and Drakakis-Smith, 1993; Rigg, 1997; Rodan et al., 1997) these changes are associated with the spread of export-oriented manufacturing and the accompanying investment from developed countries, primarily the US and Japan, and the Asian NIEs. In many Southeast Asian states, this process was and is being facilitated by national government policies to encourage such investment, such as tax-free zones and state expenditures on the development of human capital and physical infrastructure. These processes have drawn a number of the Southeast Asian countries into the global economy with a vengeance. Circumstances have progressed to such an extent that those states or regions that had initial advantages in terms of foreign investment, namely, Thailand and Malaysia, now face increasing competition from lower-cost producers, including some of their own socialist neighbours.

The Economic Crisis

A number of policy-makers and politicians argue that the cultivation of foreign investment and too much reliance on an export-oriented manufacturing strategy were the primary causes of the recent economic crisis in the region. The countries most affected by the crisis were Thailand, Indonesia, Malaysia and, to a lesser extent, the Philippines. It is more generally agreed, however, that rapid financial liberalization, with inadequate attention paid to problems of corporate governance and prudent regulation, together with poor macroeconomic management, and other policy and institutional failures, are common factors underlying the problems of these four Southeast Asian economies. In retrospect, it appears that their previous economic growth had been built, at least in part, on shaky and unsustainable foundations.

Many observers agree that the macroeconomic collapse in the region started as a crisis of confidence, with the most immediate and devastating impact being a massive flight of capital, first from Thailand and then from Indonesia, Malaysia and subsequently the Philippines. While there are many common features, the macroeconomic history of the crisis differs considerably across Southeast Asian countries. In the hardest hit countries, the loss of confidence arose from a large and rapidly increased dependence on short-term private capital flows and corresponding current account deficits (World Bank, 1998). The domestic financial sectors in all countries suffered from a variety of weaknesses that were hidden or appeared to be irrelevant in the face of rapid growth and the seemingly endless appreciation of real estate and stock market assets.

Although the latest projections from the International Monetary Fund (IMF, 25 September 2001) indicate that the region will experience average rates of growth of about 3 per cent in 2001, the two-year crisis devastated several of the most rapidly growing nations. To recap briefly, in the early part of 1997, Thailand exhausted her foreign exchange reserves through ill-considered attempts to fight off speculative attacks on the Thai currency, the *baht*. Historically, the Thai *baht* had been linked or pegged to the value of a basket of other developed economy currencies (primarily the US dollar). The country was forced to abandon the *baht*'s peg to these other currencies on 2 July, 1997. Capital flight accelerated throughout the year because of self-reinforcing negative expectations about the *baht* in turn caused corresponding collapses of the stock and property markets. This exposed weaknesses in the financial sector, leading to the eventual closure of many finance companies and a corresponding liquidity crunch throughout the economy. There was a rapid and generally unpredicted contraction of domestic demand, propelling Thailand into a major recession (TDRI, 1999).

Indonesia followed a month or two later. Initially, Indonesia seemed to be handling its situation better than Thailand. The government did not experience a long period of denial or a costly attempt to defend the currency; rather, the national government appeared to act quickly and decisively to address problems in the financial sector. A few key policy misjudgments, together with growing political and social uncertainties associated with the health and political plans of President Suharto, contributed to the currency collapse and caused an economic recession much more severe than in Thailand. While Thailand's currency, at one point, fell to 45 per cent of its pre-crisis value, Indonesia's dropped by as much as 85 per cent. Indonesia's economic growth rate in 1998 was about -15 per cent while Thailand's was -8 per cent (Table 1.5). Similarly, Indonesia experienced a much higher inflation rate over the first two years of the crisis than other Southeast Asian countries.

Table 1.5 Impact of Economic Crisis on Southeast Asian States

Country	Greatest % change in currency value	1998 Rate of growth (%)	1999 Rate of growth (%)	Inflation rate (1998–99) (%)	Projected rate of growth 2000
Cambodia	18.51 (4/97–5/97)	1.0	n/a	14.78	n/a
Indonesia	123.11 (12/9 –1.98)	−13.68	−0.79	59.57	2.57
Laos	53.43 (1/98–2/98)	4.0	n/a	81.02	n/a
Malaysia	17.36 (12/97–1/98)	−6.7	2.4	5.28	6.5
Myanmar	4.11 (12/97–1/98)	7.0	n/a	10	n/a
Philippines	15.35 (11/97–12/97)	−0.54	2.2	9.72	3.5
Singapore	5.26 (11/97–12/97)	0.35	4.5	−0.27	5
Thailand	28.95 (6/97–7/97)	−9.4	4	8.1	4
Vietnam	6.48 (7/98–8/98)	3.5	3.5	n/a	4.5

Source: IMF World Economic Outlook Database, September 1999. Available at:
    www.imf.org/external/pubs/ft/weo/1999/02/data/index.htm

Leinbach (2000), among others, argues that Indonesia's inability to create an effective response to the crisis was due to an unsustainable process of political and economic activity. In essence, many of the companies that invested and made financial decisions based on a pattern of historical economic expansion had little but their political connections to the Suharto regime by way of financial guarantees as to their fiscal viability. Once the Indonesian *rupiah* began to depreciate and exports did not increase as expected, the economy began to collapse. The private sector postponed or cancelled planned investments that resulted in the lay-off of millions of workers and the government was forced to take over the financial sector of the economy. Apparently, Indonesia's banks had been borrowing excessively outside of the country to finance risky investments with little or no central bank supervision. The deeper implications of the inability of the government to manage the macroeconomy allowed long-hidden doubts about the way the state was being governed to surface, and become the subject of public debate and protest. The enormous uncertainty about the Indonesian state's abilities to govern the country helped to marshal support for reform in the economic sphere and to attack the privileged position of President Suharto's family and corrupt business associations (McLeod, 1998). One can argue that for many in Indonesia, the crisis may ultimately prove to be the most important event in sparking important political change and reform.

Neither Malaysia nor the Philippines, and certainly not Vietnam, Laos, Myanmar or Cambodia, had the levels of accumulated short-term debt liabilities of Indonesia or Thailand. As such, none of these countries suffered the financial downturns of Indonesia and Thailand. While the Philippines experienced much of the general contagion effects of the spreading crisis, it weathered the crisis relatively well. Some researchers argue that the Philippines had moved away from dangerous links between government and financial institutions since the economically painful Marcos era. In addition, the Philippines experienced a financial system crisis between 1983 and 1985 that resulted in more stable and more prudent central bank supervision (Lim, 1998). Although Malaysia did not have as serious a short-term debt problem as its neighbours, it did suffer from financial sector weaknesses arising from questionable and often politically motivated lending policies. This, combined with growing and increasingly open divisions within the government about the desirability of Malaysia's exposure to global financial forces, diminished market confidence and led to capital outflows (Jomo, 1998). Directly flouting 'IMF orthodoxy' Malaysia chose to insulate itself from global financial markets through the imposition of direct capital controls.

Domestic policies and politics clearly played an important role in the size of the aggregate macroeconomic shocks arising from the region's crisis. While a great deal of resources have been spent trying to understand and explain the causes of the crisis, relatively little research has focused on the impacts of the recession on society writ large. An important impact, from the perspective of urban quality of life, centres on the ways that the crisis affected expenditures and projects in the sphere of environmental management. It is very likely that most national governments cut back the amount of resources allocated to basic environmental programmes such as water treatment, sanitation services and air pollution control as part of the austerity measures imposed on Southeast Asian countries by the IMF and the World Bank. Unfortunately, few of these countries devoted enough money to protect degraded environments when their economies were growing at record levels. The crisis has exacerbated this problem in that many of the most environmentally vulnerable cities and their populations now have fewer resources to devote to the environment than before. We return to this issue of environmental management in Southeast Asia in the following chapter.

Urban Migration and Population Growth

Most Southeast Asian nations, except for Singapore and Brunei Darussalam, are still generally described as rural in nature with a relatively low proportion of the total population living in urban settlements (that is, between 10 and 50 per cent as of 1995). Since 1980, however, concomitant with the increased emphasis on manufactured exports and strong economic growth, urbanization rates in many Southeast Asian states have risen considerably. Urban growth rates in Southeast Asia have averaged over 4 per cent in all of the NIEs and Laos since 1985 with urbanization rates in Burma and Cambodia not far behind. The primary causes of this increase include the effect of rural to urban migration, relatively high natural birthrates in urban areas and changes in administrative boundaries that define metropolitan regions (Table 1.6 shows urbanization and recent urban growth rates in Southeast Asian countries).

Table 1.6 Urbanization and Urban Growth Rates

Country	Percentage of population urbanized 1999	Average annual growth rate of urban population (%)	
		1970–90	1990–99
Cambodia	16	1.5	5.0
Indonesia	40	5.0	4.5
Laos	23	5.3	5.4
Malaysia	57	4.5	3.7
Myanmar	27	2.4	2.3
Philippines	58	4.4	4.1
Singapore	100	1.9	1.7
Thailand	21	3.9	2.4
Vietnam	20	2.6	1.8

Source: Bellamy, 2001.

Migrants are drawn to Southeast Asian cities both by traditional pull factors such as better educational and employment opportunities as well as by the push factors of deteriorating conditions and limited options for wage earning due to technological improvements in agricultural production, poverty, landlessness and sometimes political instability. Political upheaval or traumatic events, such as the Vietnam War or the Khmer Rouge period of Cambodian history as well as the more recent removal of Suharto in Indonesia, are associated with increased urban migration. Often, these conditions are exacerbated by the high rates of natural increase in rural areas as health conditions (particularly prenatal and infant care) improve without the widespread acceptance or use of birth control.

In the cities of Southeast Asia, rates of natural increase are also quite high, on a global scale, primarily because a high percentage of first generation migrants live in urban areas and bring with them their traditional reliance or desire for large families. While the rates vary, natural increase accounts for approximately 50 per cent of the growth in urban population in Southeast Asian countries. In the 1990s, natural increase explained about 60 per cent of the urban growth in the Philippines, 45 per cent in Malaysia and about 35 per cent in Thailand. Most of the residents of urban areas in the Southeast Asia are very to somewhat poor (Table 1.7). Urban poor residents often live in large informal areas of metropolitan areas having both permanent, long-term residents and recent migrants who live in structures which appear temporary but have often existed in similar form for many years.

Vietnam

Source: Kate Swanson

Table 1.7 Urban, Rural and National Poverty

Country	Survey year	Population below national poverty line		
		Rural (%)	Urban (%)	National (%)
Cambodia	1997	40.1	21.1	36.1
Indonesia	1999	n/a	n/a	27.1
Laos	1993	53.0	24.0	46.1
Malaysia	1989	n/a	n/a	15.5
Myanmar	n/a	n/a	n/a	n/a
Philippines	1997	50.7	21.5	36.8
Singapore	n/a	n/a	n/a	n/a
Thailand	1992	15.5	10.2	13.1
Vietnam	1993	57.2	25.9	50.9

Source: World Bank, 2001.

Unfortunately, the rapid pace of urbanization in Southeast Asian nations has not been accompanied by investments in urban infrastructure so that most, if not all, basic services provided to urban residents are woefully inadequate.

Water, sanitation and electricity are often sporadically supplied or available to households in urban areas while full-time employment and adequate housing is generally out of the reach of millions of urban dwellers (Table 1.8). This lack of access to public services is related to these households experiencing insecure land tenure (for example, squatting) because they cannot afford to rent or buy in the formal housing market.

Table 1.8 Urban Households in Formal Settlements with Access to Services in 1998

Country	City	Potable water[1] (%)	Sewerage connection (%)	Electricity (%)	Telephone (%)
Cambodia	Phnom Penh	45	75	76	40
Indonesia	Jakarta	50	65	99	n/a
	Semarang	34	n/a	85	n/a
	Surabaya	41	56	89	71
Laos	Vientiane	87	n/a	100	87
Malaysia	Penang	99	n/a	100	98
Myanmar	Yangon	78	81	85	17
Philippines	Cebu	41	92	80	25
Singapore	Singapore	100	100	100	100
Thailand	Bangkok	99	100	100	60
	Chiang Mai	95	60	100	75
Vietnam[2]	..	47	43	n/a	n/a

[1] Access to potable water is defined as access to safe drinking water within 200 metres of the dwelling
[2] Data reflects all urban centres in Vietnam between 1990 and 1997 (WRI, 2001)
Sources: World Bank, 2001; World Resources Institute, 2001.

The lack of provision of public services, such as water, sanitation and solid waste disposal, has severe detrimental impacts on the environment and on public health. For example, household- and individual-level research on squatters and informal communities has shown that human waste in Southeast Asian cities flows directly into open canals or is wrapped in newspaper and thrown into canals or other waterways, water is obtained from public standpipes (either using vendors or transported by families to their shelter), electricity is accessed illegally through nearby utility lines and garbage is burned or left to fester on unused plots once it has been scavenged for recyclable or re-usable materials.[4]

[4] See, for example, Jellinek (1991), Goss (1997), Daniere and Takahashi (1999a).

Scavenging is a major source of employment for many citizens and provides a legal income as well as efficient recycling of waste. In Hanoi, for example, the informal scavenging sector is estimated to process about one-third of all the city's waste and employs more than 6,000 people (Hiebert, 1996). In the case of water and electricity, however, the poor often pay inflated prices charged by private, generally middle-class residents who have their own connections to municipally provided services.

Vietnam

Source: Waste-Econ Project

There are conflicting interpretations about the implication of the lack of public services available to a large, and growing, population of urban dwellers in Southeast Asia. Earlier research (for example, Dwyer, 1968 and McGee, 1967) has argued that the inability of Southeast Asian cities to absorb and/or adequately provide a modern way of life to their residents is evidence that there are too many residents in urban areas. In contrast, more recent scholarship (such as Brennan, 1995) has indicated that the lack of provision of basic services in large cities of Southeast Asia is primarily caused by limited interest on the part of governments and civil servants to capture the energy and abilities of even the poorest urban residents and to use these energies in a creative and innovative manner.

Despite the critical infrastructure and environmental problems existing in Southeast Asian cities, the overall experience of day-to-day living in urban areas of Southeast Asia is quite remarkable and even inspiring. In the face of poverty and land tenure insecurity and the difficulties involved in working and living in these cities, poor communities do not appear to exhibit widespread depression or bitterness, but are simply intent on shaping their lives in the same ways as their more fortunate or elite neighbours.

Thailand

Source: Kate Heming

The ability of the poor to protect themselves from urban environmental pollution, however, is limited and the consequences of chronic unabated exposure to pollution and its reinforcement of poverty is just being uncovered.

The Nature and Impact of Urban Contamination

Asia already has what many experts consider the filthiest air in the world, the dirtiest water, the most rapid destruction of agricultural land, as well as, of course, the most worrisome levels of overfishing and the fastest-disappearing coral reefs and forests.[5] Dua and Esty (1997), for example, state that the worst pollution in the world is unequivocally in Asia. Published data indicate that while the pollution levels in Chinese cities are astonishingly high they are followed closely by almost every other major city in Southeast Asia including Bangkok, Manila and Jakarta. Asia's problems, in general, and Southeast Asia's specifically, are particularly severe because pollution tends to reflect two fundamental forces – industrialization and increasingly large and concentrated urban populations. The filthiest smoke and water are often created in the early stages of industrialization, much of Asia's current level of development, and the region's population is dense and growing rapidly. The problems are acute in the new mega-cities, those with more than 10 million people. Asia has nine of the world's fourteen mega-cities, including the largest, Tokyo, and the fastest growing, Dhaka in Bangladesh. Thus, while the cities of Southeast Asia do not represent the extreme of any situation, there are many cities throughout the globe with much in common to the Southeast Asian cities described here.

Air Pollution

It is widely accepted that air pollution is a serious problem in most large cities of Southeast Asia.[6] Conservative estimates suggest that hundreds of millions of Southeast Asians inhale extremely polluted air. As Table 1.9 indicates, air pollution in at least three Southeast Asia cities exceeds World Health Organization (WHO) guidelines for one or more contaminant. Furthermore, air pollution emissions across Asia also seem to be deteriorating in many Southeast Asia countries. Particulate and ozone levels, in particular, are increasing quite rapidly as are rates of sulphur dioxide and nitrous oxide emissions.

---

[5] Southeast Asia has the dubious distinction of having the fastest rate of deforestation in the world, estimated at 1.2 per cent loss per year (FAO, 1998).

[6] However, air pollution in rural areas can also be horrendous as was true during the summer of 1997 when raging forest fires in Indonesia caused severe smoke pollution throughout much of region. While forest fires are generally thought to be an anomaly, urban air pollution in the cities of Bangkok, Jakarta and Manila is chronic and dangerous.

Thailand
Source: Kate Heming

Table 1.9 Urban Pollution in Selected Southeast Asian Cities, 1999

City	SO^2	SPM	NO^2	Pb
Bangkok, Thailand	x	xxx	x	xx
Jakarta, Indonesia	x	xxx	x	xx
Kuala Lumpur, Malaysia	x	xx	–	–
Manila, Philippines	x	xxx	–	xx

x   Low pollution   xx   Moderate to heavy pollution
xxx   Serious pollution   –   Insufficient data

Notes:
(1) SO2 = sulphur dioxide; SPM = suspended particulate matter; NO2 = nitrogen dioxide; Pb = lead.
(2) "Serious" refers to situations in which WHO guidelines are exceeded by more than a factor of two; "moderate to heavy" refers to situations in which WHO guidelines are exceeded by up to a factor of two (short-term guidelines exceeded on a regular basis at certain locations); and "low" means that WHO guidelines are normally met, but that short-term guidelines may be exceeded occasionally.

Sources: UNEP, 1999; Seager, 1995.

As elsewhere, a great proportion of the increase in air pollution is attributable to the burning of fossil fuels as part of an increase in industrial activity, an unprecedented increase in the rate of private car ownership and use, and congestion. Data on car ownership in Southeast Asia is particularly compelling. The number of automobiles registered in Thailand, most of which are located in Bangkok, grew from 2 million in 1990 to 7.5 million in 1997, while car ownership in Vietnam has grown by a similar percentage in the last five years (AAMA, 1998). Air pollution is aggravated by the extremely poor quality of fuels used in developing countries, such as unwashed and soft coal or dirty and/or leaded gasoline, as well as by the inefficient engines and machines consuming the fuel.

High levels of air pollution have overwhelming impacts on public health. Respiratory diseases such as pneumonia, bronchitis, asthma and emphysema are at record levels in several Southeast Asian cities. Particulate pollution, either on its own or in combination with sulphur dioxide, has led to an enormous burden of ill health and is estimated to have caused 4-5 million new cases of chronic bronchitis each year (World Bank, 1999). The World Bank has published estimates suggesting that between 2 and 5 per cent of all deaths in these cities could be averted if particulate levels in developing country cities met WHO standards (World Bank, 1998a).

In Jakarta, total suspended particulates were estimated to average 271 micrograms per cubic metre in 1995, far exceeding WHO annual mean guidelines of 90 micrograms per cubic metre (World Bank, 1999). This level of particulates contributed to an estimated 5,000 deaths per year in that city alone (Bartone, 1994). Both Bangkok and Manila reported total suspended particulate levels of over 200, and while this does not approach levels found in Delhi (415) or in some Chinese cities, the levels are very troubling. There is no doubt that the urban poor, that is those who spend most of their lives on the street, and have limited access to health care, are the most likely to suffer the detrimental health effects of polluted air.

Water and Sanitation

Water issues, as in many places around the globe, can be characterized in two ways: quantity and quality. The problem of water quality in Southeast Asia, or the lack of clean and potable water, has been widely discussed by researchers and policy-makers. Given the rapid economic growth experienced in Southeast Asia over the last decade, and assuming a trickle-down effect, quality of life should have improved for all citizens in countries such as Malaysia, Indonesia and Thailand. Many urban residents, however, now face more polluted living conditions in the rapidly expanding economy than they did ten years ago. As Table 1.8 suggests, inadequate access to water and sanitation in many urban areas of Southeast Asia is a major problem.

Water in Southeast Asian cities is polluted by untreated sewage being dumped into rivers and waterways, eventually making its way into large waterways and groundwater aquifers. The ongoing pollution and lack of water treatment has had varying public health impacts. Contaminated water

Manila, Philippines

Source: www.makingcitieswork.org/
photolib/photos/maqnila2.

Thailand

Source: Kate Heming

spreads disease and, over time, can reduce the human immune system's capacity to withstand epidemics of communicable diseases. Pathogens and other bacteria are released into open water and cause a variety of diseases ranging from gastroenteritis to the deadly cholera or typhoid. In Bangkok, for example, more than 10,000 tons of raw sewage flows into the city's *klongs* (canals) each year. Fecal coliform counts in Bangkok often exceed WHO recommended guidelines for safe drinking and bathing water by several thousand. Families living in squatter conditions along the *klongs* depend on this source for daily household tasks, such as cooking and bathing. Such exposure to raw sewage every day poses great public health risks. Consequently, poor public health practices, such as the prevalence of untreated sewage in a city of several millions or more, can lead to decreased labour productivity as well as, of course, a decline in quality of life.

To address this expanding problem, researchers and policy-makers have worked to identify the point sources of water pollution in many cities across Southeast Asia. While much of the sewage released into the waterways in Southeast Asian cities is household waste, an increasing proportion of urban wastes are composed of solid and toxic wastes from industry and large-scale agriculture. According to the International Institute for Environment and Development (IIED, 2001), concentrations of chromium, magnesium and cadmium in some Indonesian cities were more than 100 times the limit set by the WHO. Exposure to these toxic substances not only harms residents who come into regular contact with the wastes but also causes long-term damage to agriculture and livestock such as increasing forms of cancer and destroying species of grains and other vegetables.

A typical example of this occurred in January 2001 when a major oil spill in the Marikina River, north of Manila, caused substantial damage to the aquatic life and prevented household use of the river for drinking and bathing. A boiler of the soft drink giant Coca-Cola Bottlers Philippines Inc. (CCBPI) Antipolo Plant (which ceased operation in December 1999) accidentally spilled oil into the river as it was being removed from the facility. Fish and shellfish harvested in the river and other nearby tributaries were significantly affected by the oil spill and the Philippine Department of Health (DOH) warned at the time that, 'Health effects that may occur due to repeated exposure to the river waters contaminated with oil may result in dermatitis, irritation/corrosion to the eyes, gingival and mucous membranes' (6 2001). The DOH virtually declared the Marikina River dead, and banned residents from using water from the river for all purposes despite the assurance of CCBPI that the oil spill was adequately contained.

Indonesia
Source: Marlene Doyle

While water quality has been the focus for much research on urban environmental issues in developing countries, of equal importance is an adequate supply of water, the quantity issue. The scarcity of water is of great concern in many Southeast Asian cities. Although the region as a whole is water-rich because of abundant annual rainfall and high water tables, many of the urban areas are not equipped to provide water for the number of people who now live within their water catchment basins. Jakarta, for example, receives a great deal of annual rainfall (about 80 inches a year) yet the water supply available to city residents is inadequate. The municipal system does not extend to certain areas and/or only reaches to a few locations in some of the poorest parts of the city. Residents are obligated to queue for piped water for many hours, or to purchase water at very high per litre cost from water vendors who hold concessions for certain public water pipes (Crane and Daniere, 1996).

Even in smaller cities, such as Hat Yai in Thailand, urban water supply is a matter of great concern due to the rapid depletion and contamination of groundwater aquifers. Lawrence et al. (2000) recently found that due to widespread contamination of shallow groundwater aquifers in Hat Yai (Thailand's third-largest city), the majority of the potable water supply is obtained from groundwater in deeper semi-confined aquifers 30 to 50 metres below the surface. However, the contaminated shallow groundwater is one of the sources of recharge water for the deeper aquifers and is apparently causing the water in the aquifers themselves to become contaminated, albeit very slowly. Unfortunately, because of the extended time necessary for better quality water to flush through the aquifer, the water quality even of these deeper aquifers will be compromised for as long as thirty years even if the groundwater were fully treated today. Clearly, this impacts upon the long-term supply of potable water for city residents.

The problem throughout cities in Southeast Asia is compounded by the high rate of leakage in municipal water pipe systems. Up to two-thirds of municipal piped water is lost through leakage (leaky pipes and connections, and deliberate tapping into the system by residents) in many Southeast Asian cities. Although access has improved somewhat in recent years in Jakarta, water consumption is still lower than WHO standards suggest are appropriate, and water remains quite expensive and inconvenient to obtain (UNESCAP, 1998). Furthermore, the groundwater in and around Jakarta has become both contaminated and salty due to salt-water intrusion from the ocean. Lack of water combined with poor-quality or contaminated water is a dangerous and unhealthy combination experienced in many Southeast Asian urban settlements.

Indonesia
Source: Marlene Doyle

To address the problems of water quantity and quality requires the construction of infrastructure and/or the development of water delivery systems. However, because of the lack of financial capacity for most Southeast Asian governments, the construction of new infrastructure and/or the development of new systems of delivery require the identification of financing sources. A large part of improving water provision in Southeast Asian cities involves finding the resources to pay for construction and maintenance of distribution and treatment facilities. One strategy focuses on water pricing as a means to finance new infrastructure and systems. However, water pricing is a major problem in Southeast Asian cities. Many Southeast Asians believe that clean water should be free because it is such an important good and because, in many places, apparently so abundant. Even many community leaders and politicians have not begun to appreciate that water provision and treatment costs money, and that providing clean water to a large and polluted city costs a great deal of money. As a result, pricing schemes that attempt to recoup the costs for providing water to city residents are often rejected, or unevenly applied with some areas paying more for water than others (Briscoe, 1999). Educating the population, as well as finding the resources to fund this work, are critical first steps on the way to developing appropriate solutions.

Solid and Hazardous Waste

Solid and hazardous waste generation is a widespread problem in rapidly changing areas of Southeast Asia as they are in most industrializing parts of the world. Most Southeast Asian cities, with the exception of Singapore, are

experiencing rapid expansion in solid waste (for example, household waste and other rubbish). It is common to find vacant or undeveloped plots of land filled with the neighbourhood rubbish, swarming with rats and other pests in Jakarta, Bangkok and Manila. Obviously, large cities generate a great deal of solid waste that needs to be collected and disposed of in a safe and environmentally sound manner. Unfortunately, the management of a large volume of waste such as that created in a city of 5 or 10 million people is complicated and relatively expensive.

Vietnam
Source: Molly Davidson-Welling

Southeast Asian cities appear not to have the capacity to dispose of waste in safe and environmentally sensitive ways. In many localities, waste collec-tion is inadequate and the waste is disposed of in a dangerous or inappropri-ate manner. Most of the municipalities in Thailand, for example, are only able to collect a portion, often less than 50 per cent, of the waste generated by house-holds and commercial establishments in the municipality. Even when household and other solid waste is collected, it is often incinerated (which contributes to air pollution) or dumped in unlined landfills. The unlined nature of the landfills means

Vietnam
Source: Molly Davidson-Welling

that surface and groundwater contamination is likely. Solid waste collects in front of drainage channels and contributes to the spread of disease and infection through land and groundwater contamination. Local workers and proximate residents are exposed to a variety of vermin-generated disease, potentially contaminated ground-water and treacherous working conditions. The disastrous avalanche of garbage in the Patayas dump in Manila which killed more than 200 scavengers in 1999, lead-ing to its subsequent closure and a lack of space to dispose of rubbish in Manila, is an internationally known horror story. Improved management of solid waste is a matter of better collection and disposal that would require investment in infrastruc-ture and public service delivery. Taxpayers and residents of Southeast Asian cities are not used to paying a lot, if anything, for this service and municipalities are often too poor to fund comprehensive rubbish

Vietnam
Source: Waste-Econ Project

removal and disposal services. We return to the challenges of providing solid waste public services in the following chapter.

Hazardous wastes, such as chemical solvents, have also become very prob-lematic for Southeast Asian cities. Unfortunately, most Southeast Asian countries are at the stage of development that appears to be associated with large expansions in the level of toxic wastes generated per person. The challenge in Asian nations has been the rapid expansion of hazardous wastes (generated both at home and abroad) without concomitant means for safe containment. A country such as India, for example, produces almost 40 million tons of hazardous waste each year which, even on a per capita basis, is many times higher than that produced in Japan or South Korea (Brooks, 1998). Many policy-makers in the Philippines are particularly concerned with what to do about the hazardous waste that is now collecting on the streets of the city. A recent article in the *Manila Times* noted that the problem is not only the regular household waste (now becoming permanent fixtures in Manila's neighbourhoods), but it is also the specially toxic and hazardous wastes from hospitals that health officials are particularly worried about (Laurel, 2001). The country's Health Secretary, Secretary Romualdez, recently announced his intention to review the country's Clean Air Act's ban on waste incinerators since the use of incinerators is thought to be a possible solution to the problem.

In some cases, large and/or multinational corporations in Southeast Asia appear to be improving their management of both solid and hazardous waste materials (Bryant, 1997). Small or medium-sized firms, however, lack the resources and capacities to install appropriate technologies for handling their hazardous waste. Perhaps even more distressing is the documented movement of hazardous waste from developed countries to developing countries that generally have fewer and more poorly enforced laws to regulate the disposal of hazardous wastes. Most developed countries, such as the US, Japan and Singapore, eventually develop and adopt technologies that permit increased production with significantly lower levels of toxic by-product created per capita, and develop strategies for storing and containing hazardous wastes, though siting such storage and containment facilities has also been highly problematic (Lake, 1993). Alternatively developed countries, to overcome barriers and obstacles at home, at times look to developing countries as repositories of waste because of less stringent environmental regulations. Despite a number of international efforts to correct for differences in environmental regulations (such as the Basel Convention of 1989, which attempts to regulate trade in hazardous wastes), the widespread activities of black market dealers in hazardous waste is well-documented (Worldwatch Institute, 1999). These dealers facilitate the movement of hazardous waste from developed countries where costs for disposal are high (due to strict regulations and community opposition to hazardous waste facilities) to developing countries where disposal costs are much lower (due to lax enforcement, a lack of regulation and a lack of awareness on the part of the populace). Needless to say, most Southeast Asian countries are more likely to receive hazardous wastes from other places as well as to experience contamination from hazardous wastes generated within their borders than other more developed countries in Europe and North America (Krueger, 1999). Concomitantly, except for perhaps the factory workers themselves, the poorest residents of areas where hazardous wastes accumulate, generally urban areas close to dumping sites or waterways, are the most vulnerable populations in Southeast Asia in terms of exposure to hazardous wastes.

A recent example from what is perhaps the worst slum in the world is located on the edge of Phnom Penh, Cambodia, on a muddy slope leading to the Bassac River. Along the wooden planks that serve as walkways over the putrid mixture of mud and human wastes, nearly every woman there acknowledges having lost a child or two over the last few years (Kristoff, 1997). The tens of thousands of people living in this slum drink the river water that is foully polluted with domestic wastes from households as well as a myriad of industrial effluents, including hazardous wastes. While the circumstances of the Phnom Penh slum are extreme, severe contamination of the living environment by domestic and industrial wastes (both hazardous and non-toxic) is common in urban areas of Southeast Asia.

How have Southeast Asian governments, at the local, regional and national levels, dealt with this urban environment? What are the institutional, political and social challenges to designing and implementing environmental management policies, and enforcing regulations meant to address these critical issues? The next chapter takes up these questions and assesses the varying measures and strategies used by states across Southeast Asia in coping with urban environmental degradation.

Chapter 2

Environmental Management Systems in Southeast Asia

Environmental management in Southeast Asian cities is fraught with significant challenges. The widespread combination of policy strategies stressing limited urban environmental protection and prioritized economic growth have meant that governments at many levels engage in minimal regulation and enforcement with respect to improving and sustaining urban environmental quality, even where policy choices in other spheres, such as economic development, have significant negative effects on pollution and resource sustainability. There are several primary reasons that governments in Southeast Asia do not prioritize urban environmental quality as a policy issue. Environmental policy development and enforcement are extremely complicated and local, regional and even national bureaucracies lack the capacity or means to adequately perform regulatory functions. In addition, policies established in important economic sectors (generally manufacturing and resource extractive-related sectors such as agriculture, energy and transportation) often have important unanticipated yet related deleterious effects on urban environmental quality (Dua and Esty, 1997). Finally, Southeast Asian governments frequently give priority to accelerating economic growth in lieu of a policy focus on quality of life or environmental protection.

While policies related to infrastructure and economic growth are centrally important drivers of environmental degradation, this chapter is primarily concerned with how the state's form and function lead to particular policy outcomes and how local communities are affected by and attempt to address environmental problems and policies. The concern here is that individuals, households and neighbourhoods are being asked to absorb an ever-increasing level of environmental pollution and to take more responsibility for mitigating that contamination. At the same time, governments are not increasing the resources devoted to environmental protection efforts. Southeast Asian countries, such as Thailand, Indonesia and Malaysia, did little to protect urban environments during the sustained period of generally high economic growth rates during most of the late 1980s and until the economic crisis of 1997. Now, in response to low if not negative growth rates since the 1997 economic crash, these countries do even less to relieve urban contamination. More troubling is

27

evidence that communities are, in fact, left out of much of the decision-making and policy design process, despite being called upon to help implement environmental management policies. On a more positive note, such community participation may have emancipatory potential. As these local communities become more involved in environmental management, governance and state action have the potential to be significantly altered.

This chapter focuses on environmental management policies and practices in varying Southeast Asian contexts to illustrate how the dual practices of limited environmental management and prioritized economic growth have created contaminated communities. The purpose of this chapter is to highlight the limitations presented by state-centric environmental management, especially in Southeast Asia, to summarize recent research that has focused on community and user participation in infrastructure and water management policy design and implementation, and to set the stage for our later arguments about understanding environmental governance in Bangkok (Chapters 3, 4 and 5).

Overview of the Problems

The state-centric view of environmental management has long been the perspective expressed by researchers and policy-makers (Bryant and Wilson, 1998). Much of the public health and ecological problems in Southeast Asian countries are due to environmental policy failures that stem from the intrinsic difficulty of environmental policy-making and the lack of regulatory capacity and political will in most Southeast Asian governments. Environmental policy-making at its most basic requires numerous steps and a sophisticated knowledge of techniques and procedures. Unfortunately, most civil servants working in developing countries in Southeast Asia (and many other developing and developed nations) lack these technical skills. And even if agency staff have developed technical skills, such bureaucracies generally suffer from a lack of personnel due to insufficient financial resources (Pistor and Wellons, 1999). This shortage of human capital in turn is exacerbated by a lack of technical and scientific equipment and contributes to inappropriate or insufficient resources for monitoring and enforcement.

Obviously, these types of problems are not unique to the governments and bureaucracies of Southeast Asia. As noted in a study conducted by UNESCAP (1999), the environmental agencies in charge of water resources management in China were deemed to lack skilled staff, suffer from a deficit of infrastructure to adequately assess pollution and possess insufficient powers of enforcement. Similarly, most of the population of Greater Buenos Aires, Argentina, suffers from a lack of piped water and provision of sewers and drains despite living in one of the most cosmopolitan cities of the world (Pescuma and Guaresti, 1991). These same kinds of inadequacies – human resource short-ages, lack of sufficient monitoring capacity and so on – also obstruct urban air pollution control in many ASEAN countries (UNESCAP, 1995). Industrial pollution, as noted above, often goes entirely unregulated in most of the devel-oping ASEAN countries (Dua and Esty, 1997). In Indonesia, for example, as of 1995, there were only 500 employees assigned to monitoring and enforcing emission standards for the more than 75,000 industrial plants located in Java (Brooks, 1998).

Additionally, in many developing countries of ASEAN, the responsibility for environmental protection resides within a myriad of uncoordinated and often competing ministries and government agencies. The fragmented nature of responsibility means that mounting a sustained and comprehensive plan or programme to protect and monitor different aspects of the environment is extremely difficult if not impossible. Although some Southeast Asian coun-tries or regions within particular countries have tried to institutionally reform the structure of environmental management agencies, most efforts have met with failure. Agencies typically resist changes in responsibility and the accompanying alterations in budget with the result that little progress is made on the environmental front (O'Connor, 1994). Environmental condi-tions thus continue to deteriorate at a much faster pace than advances in technology or implementation of protective measures.

Another major impediment to improved environmental management, particularly in developing countries, is the widespread belief, promoted by state and private sector actors, that a period of environmental destruction or devastation is a necessary precursor to industrial development. In essence, policy-makers in Southeast Asian governments support or, at least, appear to espouse a 'pollute now pay later' strategy towards economic development in the short term (Dua and Esty, 1997). Some governments in the region argue that they are simply following the pattern adopted by the Japanese government that, in fact, placed overriding priority on industrial development and export promotion following the Second World War.[1] The result is that, as is now the case in much of Eastern Europe, the Japanese were left with areas of environ-mental devastation that, ultimately, cost the country and its citizens a great deal to mitigate and to restore (to the extent that this was possible).

Most environmental economists agree that there are a variety of reasons for developing countries to protect aspects of their environment at the same time as they modernize their economies. There are many instances where investing in the environment, as opposed to destroying it, results in benefits that far exceed costs. There are numerous examples of how controlling pollution at the source, which is generally relatively inexpensive, can result in increased pro-

---

[1] A number of authors note that Taiwan and South Korea, in particular, seem to have closely followed the Japanese model (Bello, 1993; Kim, 1998).

ductivity (as well as lower health costs and increased natural resources) particularly in the world's poorest countries. Unfortunately, these tend to be the same countries that often believe they cannot afford to monitor and enforce environmental regulations.

From a community-centred perspective on environmental management, the focus, priorities and strategies differ significantly from state-oriented viewpoints. Many government decision-makers weigh the short-term costs of environmental protection measures much more heavily than the long-term improvement in health and productivity. Environmental management may be perceived then as one strategy to balance the costs of development while 'managing' the local population. For example, it is likely that if most urban residents were aware of the environmental dangers posed by environmental degradation and the benefits of developing environmental policies, some might pressure their governments to enforce environmental programmes, especially given recent political upheaval and instability in many of these nations. That is, the majority of urban populations in Southeast Asian countries, and developing countries worldwide, are not truly cognizant of how the environment affects their long-term health and well-being. Often, residents have limited access to public officials and even less influence on policy design and implementation. With few exceptions, such as in Singapore, the public has little power and little accurate information regarding environmental issues.

It is obviously not in the interest of many government leaders to direct public agencies to inform residents of the dangers posed by economic policy decisions that favour development at the expense of public health and natural resource concerns. This is particularly the case in urban settings where the potential for conflict and political upheaval is great. Thus, in many localities, governments try to downplay or ignore the consequences of environmental destruction to the population at large. Efforts to educate the public regarding health measures do not address the cause of the problem nor explain what the government is doing to prevent or mitigate environmental deterioration. Without adequate public education programmes, awareness and action regarding the environment is not likely to grow or be effective (Young and Demko, 1996).

In Southeast Asian settings, then, public input into environmental decision-making is to a great extent managed by the politicians and agencies charged with the responsibility to guide the economy and safeguard the health of the population. Public input is limited and, often, ineffective due to the lack of representative governments, a controlled press and a relatively uninformed populace. As a result, the urban poor in these countries are exposed to various environmental pollutants at far higher levels than residents of cities in either more or less developed countries. At the same time, citizens in Jakarta, Manila,

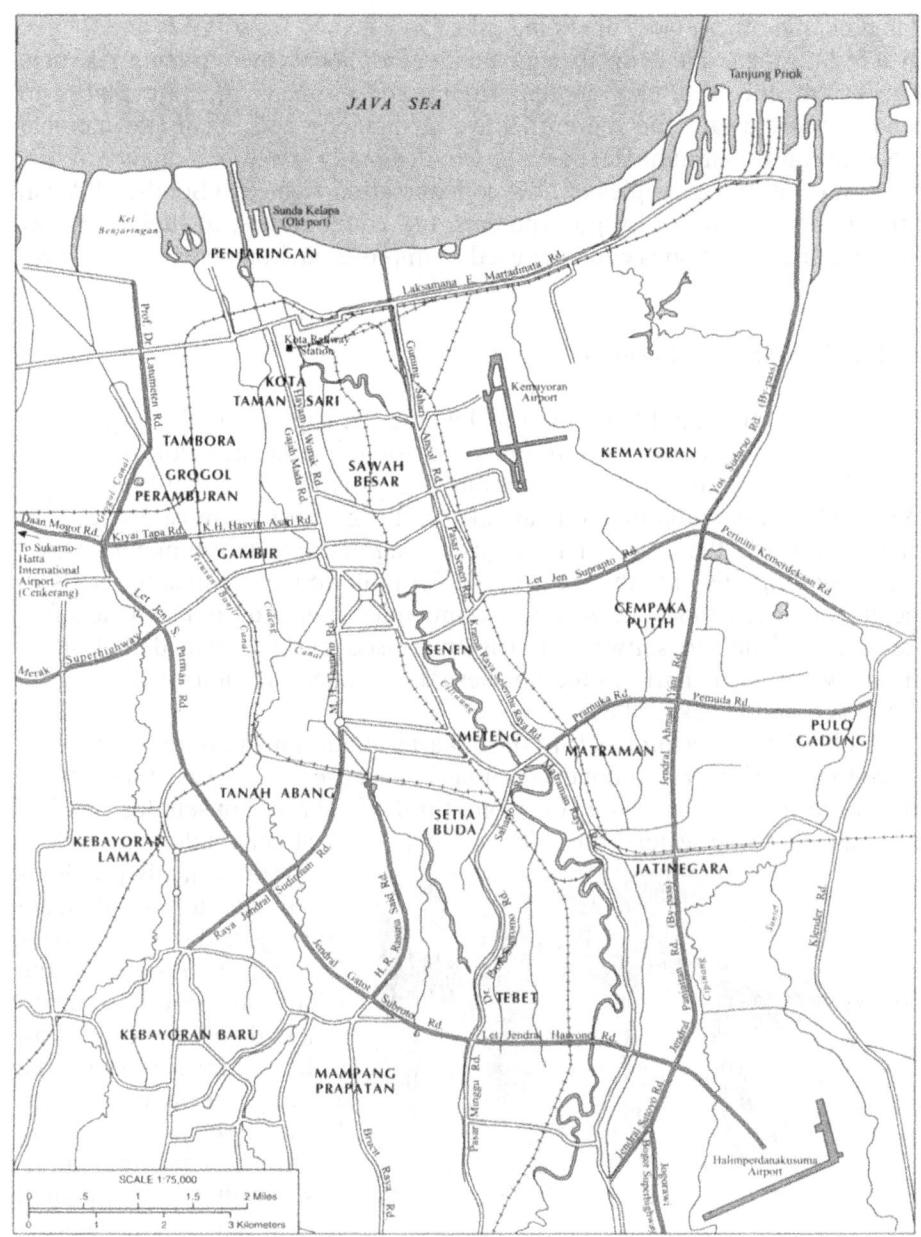

Figure 2.1 Map of Jakarta, Indonesia

Source: Ulack and Pauer, 1989.

Bangkok and other places are being asked to help pay for services and to take an active physical role in the management of their local environments. As such, the current era represents an opportunity for residents in these places to become more aware and more effective at directing both local and national economic and environmental policies, which directly affect their quality of life. We turn to several examples of how contamination in several Southeast Asian cities is being handled by governments and communities, to highlight how more inclusive strategies could be used to improve environmental conditions.

Water Provision in Jakarta

The extended metropolitan region of Jakarta, the capital and the largest city of Indonesia, including the provinces surrounding Jakarta (known as the JABOTABEK region), has a population of approximately 18 million (Hugo, 1997). This population in and near Jakarta has grown by almost 50 per cent since 1980 and there are many signs of significant environmental stress accompanying this growth (Figure 2.1). Air pollution levels in some parts of the city are often above the WHO-recommended standards and local aquifers are salinated due to saltwater intrusion caused by excessive depletion of groundwater particularly in the northern and western sections of the region (Leinbach and Ulack, 2000).

Much of the housing stock in the Jakarta region ranks poorly even compared to most shelter in Indonesia primarily because there is very poor access in Jakarta to drinking water; over one-third of Jakarta households need to buy their drinking water from relatively expensive street vendors (Struyk et

al., 1990). Equally problematic is the lack of adequate drainage and sewage collection and treatment in the Jakarta metropolitan region as less than 5 per cent of all residents are served by the existing sewage system. Jakarta's rivers, canals and shallow groundwater aquifers are often contaminated as a result of the lack of wastewater treatment and defective septic tank systems.[2]

Surakarta, Indonesia

Source: Http://images.umdl.umich.edu/i/indonesian/
(Tommy Marsudi Utomo, University of Michigan)

---

[2] Much of this discussion relies on Crane et al. (1997) and Crane and Daniere (1996). Both these papers contain a full discussion and analysis of the water supply and demand situation in urban slums of Jakarta in 1995.

More than 40 per cent of Jakarta's population essentially depends on groundwater for their drinking water, but the quality of the groundwater is very poor due to contamination. Furthermore, the groundwater table is saline several kilometres from the coast, with well water in some sections of the city quite undrinkable.

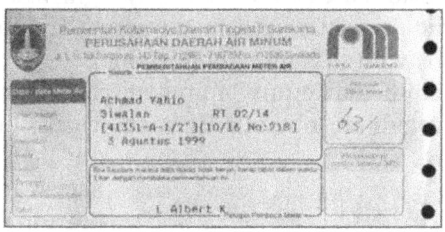

Surakarta, Indonesia

Source: Http://images. umdl.umich.edu/i/ indonesian/

Surakarta, Indonesia

Source: Http://images.umdl.umich.edu/i/ indonesian/

The other main source of water in Jakarta is the municipal water authority. As in most cities of the developing world, water is supplied by both public and private mechanisms including the most convenient mode, in-house piped connections. Only about 20 per cent of the homes within Jakarta have such a connection, which is why so many of the residents of Jakarta continue to rely on potentially contaminated and salinated wells. Residents in the north and west of the region, where the groundwater is particularly compromised, must

Indonesia

Source: Marlene Doyle

purchase their water from public hydrants (a form of private standpipe) or street vendors. The vendors acquire their water from hydrants that are connected by pipe to the municipal distribution system and transport water in metal jerrycans or large plastic containers using handcarts from door to door or lane to lane. Vendors empty their containers into a household's water barrel or reservoir upon payment.

If a household wishes to save money, it can choose to purchase hydrant water directly. In this case, a household either carries buckets or containers of water from the tap back to its home or uses a handcart rented from a hydrant operator. The service quality differs from hydrant to hydrant as there are significant differences between water pressure, reservoir size and cost of the franchise to the hydrant owner throughout the region. In general, market forces appear to determine the price per litre of water rather than official municipal guidelines. As a result, people in the northern and western areas of the city pay significantly more for their water (that is, people who use water vendors pay about 15 times the per litre price charged to households with in-house connections).

The health and equity implications of this essentially unregulated and private system are obvious. In many cases, the poorest residents in the city are the individuals who have the least access to water and pay a hefty premium for the water they consume, which forces them to consume extremely low levels of water.[3] In response to the problem, the Jakarta municipal water enterprise (Perusahaan Air Minum Jaya) embarked upon an aggressive standpipe investment programme to reduce the distance between hydrants. Such a strategy should lower water transportation costs and presumably increase competition among hydrant operators and vendors, which would ideally lower user charges. In addition, to further increase the number of suppliers and further lower costs to poorer residents, the water enterprise permitted all private households with a metered in-house connection to resell water starting in 1990.

Surakarta, Indonesia

Source: Http://images.umdl.umich.edu/i/
indonesian/

These efforts to improve the situation apparently had immediate but small effects on the price paid by households. On the one hand, households that were able to take advantage of closer hydrants or purchase water from their neighbours experienced some improvement in their access to water and the price they paid for it (Crane, 1994). On the other hand, Crane asserts that the prices did not decline to the extent some experts predicted, and there were still large price differentials several years later when a similar survey of water access and pricing tried to estimate differences in water costs among Jakarta households (Crane and Daniere, 1996). As such, while policy changes were able to address some of the problems facing poor residents regarding water prices and access, much more work needed to be done to protect existing water sources and to provide all urban residents with relatively high-quality and affordable water.

---

[3] Many researchers have documented this pattern throughout the globe. See, for example, Whittington et al. (1991) and Swyngedouw (1995).

A key step in improved provision of water supply and access lies in the consultation and participation of the users. The solution adopted by Jakarta's municipal enterprise was imposed from above: that is, they increased the supply for some users but this solution or response did not result from a process that was created or even approved of by the eventual users. Income and price data strongly suggest that poor households could afford and would purchase an individual connection or share a connection among several neighbours (Crane and Daniere, 1996). The main impediment, for many households, is the lack of a sizable amount of money available to pay for the actual cost of the connection and meter. A flexible system that responds to the financial circumstances of the users as well as their preferences might, in fact, permit the situation to be vastly improved. In addition, training or using household members to install and monitor water equipment (perhaps paying them a portion of the proceeds collected every month) could, for example, further improve the basic access to water of slum households in Jakarta.

The positive impact on water systems of public participation in the design and implementation of water and sanitation systems is well documented in recent research. Isham and Kähkönen (2000), for example, study the impact of several governmental and non-governmental organization (NGO) initiated community-based water projects in the late 1980s and early 1990s in Java, Indonesia. Their study analyses how closely a variety of these water projects followed the community-based approach, which incorporates a demand-responsive focus on what users want and what they are willing to pay.[4] In essence, they find that projects that actually allowed the users to participate in service design and, because of social conditions, were able to motivate household contributions to construction, resulted in more satisfactory and sustainable water projects with more significant and long-term health impacts than those without such user participation. In particular, in the villages of Central Java where households contributed to service construction and when household contributions to construction were monitored, communal wells and hydrants were more likely to perform effectively. Furthermore, access to a piped water service that provides water every day was associated with dramatic improvements in household health. Controlling for other household- and village-level variables, access to a piped service increased the probability of improved health by 0.29, while access to a private (in-house) connection did not have an additional positive or significant effect. The availability of water every day further increased the probability of improved health by 0.20.

The evidence assembled by Isham and Kähkönen (2000) suggests that making water services demand-responsive promotes their performance and impact: households are more likely to maintain services that match their demand. The

---

[4] For their analysis, Isham and Kähkönen use quantitative and qualitative data collected from forty-four villages in Central Java including surveys of 1100 households and 44 water committees, technical assessments of project performance, and participatory exercises in female and male focus groups.

results also indicate that one can promote demand-responsiveness of water services by involving households in the design process and by letting households, not outsiders or village leaders, make the final decision about the service type and level. Of course, it is crucial that households make an informed choice, which implies that sufficient information be provided to potential users about the cost and maintenance requirements of different service options during the design process.

The importance of community participation in the design and implementation of infrastructure projects cannot be overstated. As argued by Whittington et al. (2000), most cities in Southeast Asia need to move from a traditional conception of city-wide master planning to a focus, instead, on the needs and requirements of specific neighbourhoods, especially in terms of their demand for improved water and sanitation services. The water system in Jakarta, for example, is not able to address the needs of the urban poor (who comprise much of the city's populace) because it is not currently a technically feasible service that households believe has convenience, health and environmental benefits. Services that respond to residents' concerns will result in the construction of infrastructure in locations and in forms that they actually want, which in turn should result in high usage or connection rates, improved revenues for the supplier and better service all around. While a demand-driven approach is more time-consuming and inevitably more complicated in terms of design and implementation, it is likely to be the only approach that promises a way out of Jakarta's water management dilemma.

Whittington et al. (2000) recently conducted a case study based in the city of Semarang, Indonesia to examine consumer preferences about sanitation-related infrastructure and how willing households in poor neighbourhoods are to participate in their installation. They found that most households were not willing to pay for a connection to a sewer system despite what most engineers perceive to be an overwhelming need. However, many of the low-income households surveyed in Semarang were keenly interested in learning more about the possible sewerage and wastewater treatment technologies they were introduced to during the course of the study. It appears that a very flexible trunk system that allows neighbourhoods to opt in or out of the main sewer network might be a more reasonable approach than connecting all neighbourhoods at the same time. Furthermore, this study indicated that public education, and gradual experience with successful and affordable implementation of different sewage technologies and schemes prior to construction, may be the best way to increase household willingness to connect to and pay for a hypothetical system.

The lesson for the case of Jakarta is relatively clear. Given the existing and poorly functioning system that exists, working with *kampungs* (typical urban Indonesian neighbourhoods) to design appropriate water and sewer connections is more likely to produce much needed public revenues and better health/environmental outcomes than continuing with the *status quo*.

Manila's Rubbish Dilemma

As is true of Jakarta, Manila is the Philippines' primary locus for international trade as well as its key industrial, financial, cultural and educational centre. Metro Manila is formed by four politically separate cities (Manila, Quezon City, Pasay, Caloocan) and thirteen other urbanized municipalities (or towns) that have been restructured and integrated in the Manila Metropolitan (MMA) or the administrative region called the national capital region (NCR). Thus, the MMA consists of the original colonial core near the Pasig River, and many peripheral areas or suburbs some of which are or were separate cities before being engulfed by the mega-urban region (Figure 2.2). The urban poor are found throughout the region living in slum and squatter areas of which the most infamous is the port area known as Tondo. Squatter settlements or slums are estimated to house or provide space for about one-third of the metropolitan population (Berner, 1997).

Since 1950, the population of Manila has grown from approximately 1.5 million residents to well over 10 million in 2000. The massive urban sprawl experienced in the region has had huge implications for infrastructure provision given that investment in physical and social infrastructure including housing, water and sanitation, educational and health facilities is extremely limited and clearly inadequate. Housing is considered by many residents and planners to be a major problem because at least two-thirds of all new

Philippines

Source: www.etext.org/Politics/MIM/art/

housing constructed in the MMA since 1975 is unregulated. The informal shelter constructed by low-income residents often consists of cardboard or tin, is without access to water, sewerage or electricity, and is generally agreed to be deplorable. Housing is particularly unattainable for the urban poor in Manila because the price of land in and around the region has increased dramatically as a consequence of the legal but destructive activities of wealthy speculators and developers. The Philippines unfortunately has one of the most extreme distributions of income in Asia, so that the average income of

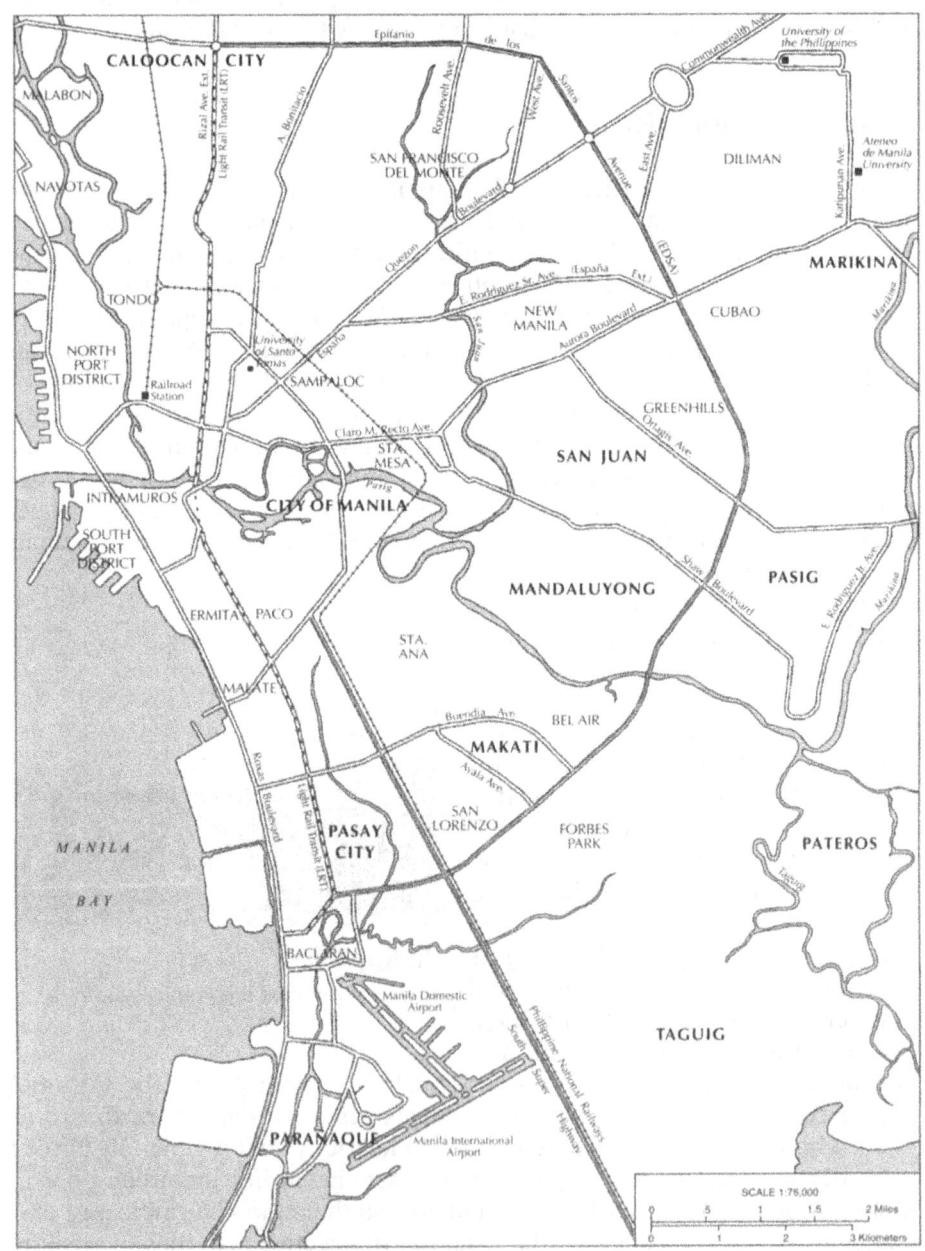

Figure 2.2 Map of Manila, the Philippines
Source: Ulack and Pauer, 1989.

the richest fifth of Filipinos is eleven times that of the poorest fifth.[5] This inequality essentially facilitates a great deal of investment on the part of wealthy developers that does much to increase land prices in Manila without necessarily increasing the incomes of most urban dwellers, particularly if the land is simply held vacant until it appreciates value enough to sell.

While housing is clearly a problem, most visitors to Manila immediately notice the highly visible amount of uncollected rubbish on streets and in the lanes, the canals and the rivers. It is similarly impossible to ignore the distressing history of Manila's rubbish dumps where it is not uncommon for many scavengers working on the dumps to lose their lives in horrendous rubbish avalanches (Mydans, 2000). The closing of Manila's rubbish dumps in recent years due to dangerous working conditions and hazardous environmental impacts has resulted in vast quantities of Manila's rubbish remaining uncollected in the streets. Although uncollected rubbish is a problem in most, if not all, Asian cities, it has reached a crisis stage in Manila.

Philippines

Source: www.etext.
org/Politics/MIM/
art/

Metro Manila, according to recent estimates, produces an estimated 6,000 tons of solid waste a year, about half of which is collected and hauled off to dump sites. The municipal authority cannot afford to collect and dispose of all the waste generated because revenues available are insufficient to service all residents and businesses. Much of the uncollected waste is left to accumulate under houses, in vacant plots of land or in canals and rivers. Individuals sometimes take pains to burn their waste or bury it near or on their dwelling site. Until 2000, there was at least one operating official dump serving the city (compared to seven dump sites in the early 1990s). However, the sites filled much more rapidly than anticipated by municipal authorities and consequently all have been closed in the recent past.

About one-third of all the waste collected in the Manila area is recycled through the efforts of families who work in the dumps to find and sell valuable waste. One of the two most infamous dump sites was known as 'Smoky Mountain' (closed in 1995) which was literally the highest point in the MMA and consisted entirely of a huge mass of solid waste picked over daily by some 12,000 scavengers. The second dump site, known as Payatas, was located in the north eastern district of Quezon City and was used for disposal of Manila's garbage for over two decades. It originally consisted of a 15-hectare open pit and apparently provided work to about 4,000 scavenger families within and

---

[5] In the next most unequal society of Southeast Asia, Indonesia, the richest 20 per cent earn, on average, only 4.5 times the income of the poorest 20 per cent (*The Economist*, 1996, p. 5).

outside the neighbourhood. On a daily basis, Payatas accepted some 1,500–2,000 tons of rubbish a day from Manila and, over time, a once verdant valley was filled and began growing into a mountain (Vincentian Missionaries, 1998). A few months before the dump was closed the rubbish covered 55 acres.

On 10 July 2000, after a week of heavy typhoon rains, a large portion of the rubbish-mountain collapsed, engulfing 400 scavenger houses that had been built around the mountain's perimeter. Despite intensive efforts by rescue crews working with heavy machinery, picks and shovels, the death toll reached 220, many of them children (Mydans, 2000). Although many scavengers have died before in Manila's slums, the Payatas tragedy made this issue highly visible at the national and international levels, and therefore acted as something of a wake up call for Manila government officials. A number of Filipino communities with similar open dumps have since started to barricade incoming dump trucks while other areas have relocated their squatter and scavenger settlements away from the dumps. In the meantime, Manila continues to generate thousands of tons of garbage a week, and for the most part, has no means to dispose of it (*The Economist*, 2000).

There are a number of problems, clearly, with the current system. One is that scavengers, who earn in the area of US$4–$5 a day, are working in dangerous conditions but, at the same time, serve a very valuable function in terms of recycling solid waste. Recycling lessens the burden on the solid waste system and provides a living to a large group of unskilled workers. At the same time, the government is apparently incapable of developing a working system of collection and disposal. Most communities in the Philippines refuse to accept a dumpsite or sanitary landfill within their boundaries. While a twenty-five-year landfill contract worth $1.1 million *pesos* a month was recently awarded to a Filipino-based consortium, apparently there will be problems in obtaining resident acceptance of the waste in Pililla, Rizal (far outside of the MMA). This proposed site is designed eventually to receive approximately one-third of all Manila's waste after the two years required for construction. In the interim, the Greater Metro Manila Solid Waste Management Committee (GMMSWMC) has awarded two-year contracts to different private waste consortia to landfill the waste on Semirara Island and in Mariveles, Bataan (another primarily rural suburb). Both localities have secured temporary restraining orders preventing the rubbish contractors from dumping rubbish in these areas, but there are indications that at least the dumpsite in Mariveles will be opened shortly (Philippine Center for Investigations, 2001).

Estrada, the President of the Philippines from 1996 to 2001, created the GMMSWMC in 1999 to develop a long-term strategy for managing Manila's solid waste. The committee is responsible for the recommendation that at least one-third of Greater Metropolitan Manila's waste be dumped into a new sanitary landfill. In addition, the Committee hopes to dispose of approximately 50 per cent of all solid waste in new state-of-the-art incinerators and recycle the remaining one-sixth of all solid waste created in the region. Unfortunately, the affordability (these options are very expensive), sustainability (reuse and recycling is not explicitly encouraged or considered) and implementation (technical and human resources are very limited) of all these plans remain significant obstacles in the resolution of Manila's waste issues. A major oversight in the development of solid waste policies in the past and, apparently, the present, is that there is little room in existing strategies for the development of more affordable and innovative technologies that can take advantage of the wealth of small-scale entrepreneurship that abounds in the Philippines and many other developing country cities (Furedy, 1997). The landfill and inciner-ation plans are both reliant on public-private alliances formed through foreign large-scale enterprises that may well prove beyond the means of the Filipino people.

A different form of public-private alliance, however, has been tried in the Philippines with some success and, for many, represents an alternative worth pursuing in many different situations. Recently several local authorities in various cities throughout the Philippines decided to draw on the professional services of private companies in designing, constructing, operating and managing their sanitary landfill while phasing out old open dump sites (Lapid, 1999). At the same time, one of the priorities of this alliance, which is funded through a loan from the World Bank and is called SWEEP, Solid Waste Ecological Enhancement Project, is to support micro-enterprise development (in particular waste dealers and itinerant buyers). The small-scale enterprises are encouraged to organize co-operatives for collection, street sweeping and recycling in densely populated areas rather than in the dumps themselves. In essence, this project represents an attempt to formalize the informal sector activities in municipal solid waste management.[6] The local authorities manage the project and retain much of the decision- and policy-making power on the design of solid waste management projects through the regulation of waste minimization and recovery. This approach has the benefit of assisting relative-ly unskilled workers to acquire skills, earn income and improve their working conditions at the same time as it prioritizes recycling, all of which incineration and using landfills alone do not achieve.

An example of how a more inclusive and sustainable approach to solid waste management can work in the Philippines is provided by an NGO known as the Metro Manila Council of Women Balikatan Movement (MMWBM). The MMWBM implemented a recycling programme by forming co-operatives of itinerant waste buyers and junk shops, providing loans, and searching for suppliers and markets. The NGO has provided the co-operatives with more bargaining power in relation to large recycling industries. In addition, the group has formed a source separation project in Metro Manila. Households in

[6] See Assad (1996) for his description of the inspirational case of the zabbaleen in Cairo, Egypt.

the San Juan municipality (the district which is home to MMWBM) are encouraged to separate their solid waste into wet and dry components. Protective equipment is provided to waste pickers and buyers who are recruited and supervised by dealers. Collection carts are funded jointly by the dealers and through small grants from the project. The buyers pay households for recyclables, relying on money received from waste dealers. MMWBM organizes the routes and schedules for buyers to collect garbage and promotes source separation through the campaigns. The council also started distributing special food and garden waste shredders to motivate schoolchildren to convert food and garden waste into compost. The local government is not involved in the NGO project (Klundert and Lardinois, 1995).

The success of the MMCWBM project is quite impressive. The NGO has provided the buyers with access to credit from government sources for working capital and for seed capital for other income-generation projects. Almost 500 dealers with more than 2,000 itinerant buyers and employees have so far joined the project, and acknowledge that better working conditions and greater public acceptance of their work has led to more and higher-quality employment. More than 200,000 households out of a total 1.5 million residents of the city are serviced with weekly collection of paper, plastics, bottles, cans, metals, car parts and batteries (Lapid, 1999). The recovery of solid waste in the San Juan municipality increased from 10 per cent in 1983 to 35 per cent in 1994. Unfortunately, the financial viability of the project is less clear. The buyers receive fixed prices for recovered materials regardless of market price fluctuations. However, regardless of the challenges faced in terms of long-term viability of the NGO, the large-scale participation by households and the large amounts of material recycled obviously contributes to a cleaner urban environment.

Air Pollution in Ho Chi Minh City

Ho Chi Minh City (formerly Saigon) is the largest city in Vietnam, with an estimated population of about 6 million in the year 2000 (including both registered and unregistered residents who have migrated without permission from outside of the Ho Chi Minh City area). It is quite large geographically (2,056 square kilometres) and densely settled (approximately 2,282 inhabitants per square kilometre) and boasts an extremely hot and humid climate (Figure 2.3). The city has grown extremely rapidly in the last fifteen years since the introduction of *doi moi* or renovation, which describes, to some extent, the Vietnamese government's loosening of trade and investment regulations for both domestic and international enterprises.

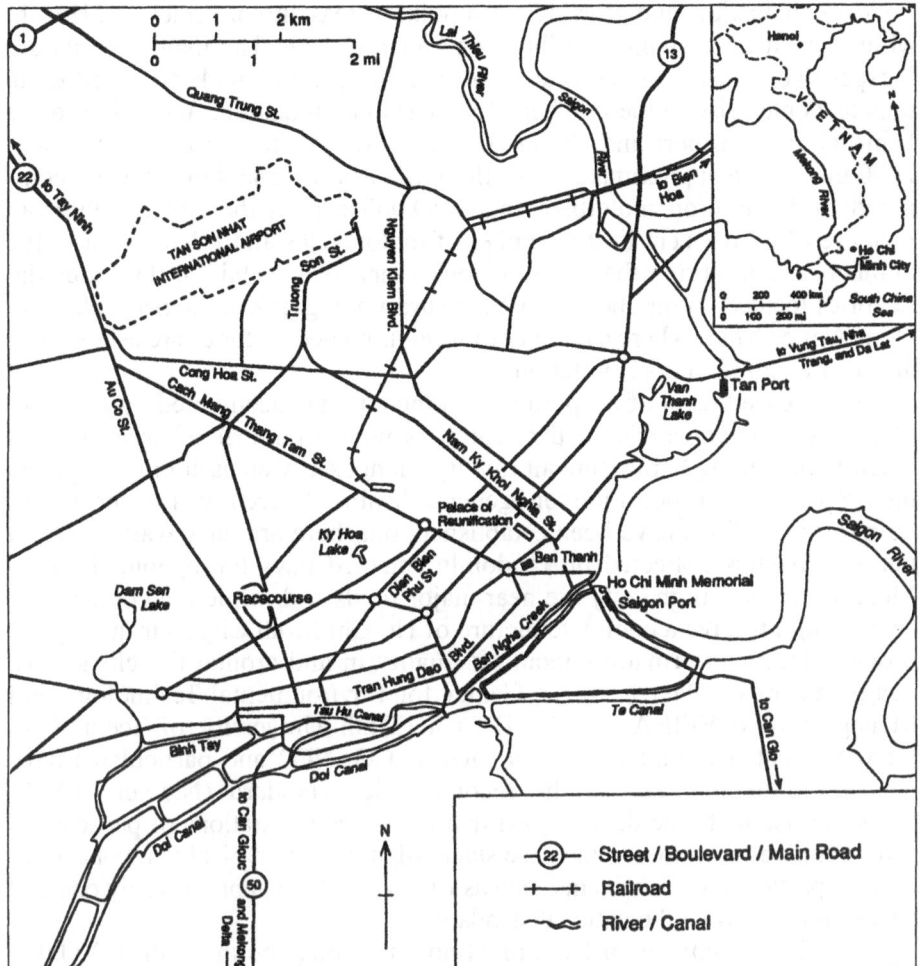

Figure 2.3 Map of Ho Chi Minh City, Vietnam

Source: Freeman (1996)

As in all bustling and growing metropolitan areas, motor vehicles in Ho Chi Minh City are the main contributors to air pollution. The main culprits are inexpensive Chinese motorcycles that have flooded the market and led to an unprecedented rise in the number of vehicles in circulation. According to the Ministry of Transport and Communication (MoTC) by the end of the year 2000 there were approximately 7 million motorcycles and 486,000 cars, compared to 5.5 million motorbikes and 450,000 cars at the end of 1999 (an increase of 27 per cent in the number of motorcycles and 8.1 per cent in the number of cars during the course of one year) (Ngoc Mai, 2001). Industrial factories located within the city and the surrounding provinces, particularly in Dong Nai province where a number of foreign-owned factories are established, also cause significant air pollution.

An increased level of air pollution in the city has been noted by a variety of different bureaucracies and researchers with a number of organizations recently beginning to measure air quality changes as well as attempt to gauge the impact of air pollution on general health. Currently, no permanent monitoring stations have been established, but there are data available for a number of sites collected on randomly selected days throughout the city. Most of the sites in the city are near major roads and, while not comprehensive enough to give a complete picture of Ho Chi Minh City's air quality, do provide decent information about air quality in and around the city's most congested areas. Data from the Center for Environmental Technology and Management (CEFINA) of the Ho Chi Minh University of Technology indicates that many air pollutants such as CO, NO_2 and particulates have concentrations that exceed the recommended standard (Nguyen, 1997). While occasionally the data suggest that hourly concentrations of particulates can be twelve times the Vietnamese standard, it is more typical that concentrations of particular contaminants such as CO or NO_2 are approximately one and a half times or twice the national standard.

In total, the vehicles in Ho Chi Minh City consume more than 200,000 tons of petrol and 190,000 tons of diesel fuel per year. Since most vehicles use leaded petrol, more than 2,200 tons of SO_2 and 25 tons of lead are emitted into the air on an annual basis. As in many parts of the world, the air pollutant concentration in Ho Chi Minh City varies depending on the season. The dry season, which lasts from October to April, sees concentrations of most pollutants that are two to three times higher than in the rainy season (Hiep Nguyen Duc, 1996).

Industrial contributions to air quality are primarily due to small factories; there are approximately 700 large industrial sources and 24,000 small factories in and around Ho Chi Minh City. The large sources are primarily located in the Nha Be and Thu Duc areas (in the suburbs) while the small enterprises (mostly family-run businesses) operate throughout the metropolitan area with the largest number of businesses to be found in Districts 5, 6, 11 and Tan Binh. Almost all of the industrial sources of air pollution are in or near residential areas. Not surprisingly, many of the technologies and much of the equipment used by the factories are old; some are as much as twenty-five years old and are clearly not fitted with any pollution control devices. In addition, many of the large foreign joint-venture establishments operate with outdated technology and cause significant environmental pollution (O'Rourke, 2001b).

Estimates generated by CENIFEA indicate that the total emission of pollutants in the Ho Chi Minh City area that are spewed into the air by industries include 30,000 tons of SO_2, 5,750 tons of NO_2 and 1,650 tons of particulates. In addition, metallurgy-based factories emit more than 4,000 tons of particulates and about 1,000 tons of CO per year. As might be expected, residents who live near some of these factories are much more likely to complain about the pollution effect of specific sources rather than the chronic long-term exposure to lead-filled air caused by thousands of slow-moving highly inefficient motorcycles. As reported in the press, residents have complained regularly about a varied selection of factories in the area, including the PS Toothpaste Company, the Miliket Noodle Company and the Southern Steel Corporation (*Vietnam Daily News*, 30 June 2001).

Few efforts, to date, document the impact on resident health of air pollution. An exception is a study conducted by the Labour Protection Unit of Ho Chi Minh City on the health effects of air pollution on traffic police officers. The study estimated that the rate of tuberculosis among traffic police is approximately three times that of the average TB rate in Vietnam. In addition, about three-quarters of the traffic police studied experienced nose, throat or ear infections in the past year with one-third of traffic police experiencing measurable hearing loss.

Overall, the measures reported by Ho Chi Minh City indicate unhealthy and declining air quality although air pollution has still not reached the levels experienced in Bangkok, Jakarta and Manila. Most observers feel this is only a matter of time, however, as the city is rapidly becoming a mega-region with all the concomitant costs and benefits of a large, growing, industrial and manufacturing urban space (Dixon and Drakakis-Smith, 1997). As such, urban residents will soon feel the dramatic health impacts including major increases in respiratory illnesses as experienced in Bangkok where one out of six people now have allergic reactions to the high level of airborne particulates (Towprayoon et al., 1997).

The response of the Vietnamese government to rising air pollution has been somewhat predictable in that it consists primarily of the introduction of laws and legislation with, as yet, relatively little enforcement, particularly in the case of motor vehicles and small factories. There are plans to improve enforcement of the pollution generated by industry particularly in industrial zones, and some permanent air monitoring environmental stations will undoubtedly be instituted in the near future. The Ho Chi Minh City People's Committee has also taken quite visible steps to move illegal houses from their location near creeks and canals to new places, ostensibly to protect water and air quality. In addition, a number of 'green days' have been initiated with the active participation of youth organizations to raise the awareness of environmental problems among the urban population.

Unfortunately, as in most Southeast Asian cities, the approach to tackling air pollution is haphazard and inadequately funded. Little is actually done to force polluters to pay for the air pollution they release into the environment. The divisions of government charged with enforcement lack human and technical capacity, and have little incentive to go after industrial polluters especially if they are foreign-owned entities bringing much need investment into the country. Furthermore, the line of authority is unclear, with generally overlapping responsibility for pollution regulation enforcement. Finally, there is no real long-term plan for solving the public health problems that are sure to intensify as motorization proceeds and the industrial base expands.

Unlike sewage installation, water provision or solid waste collection, air pollution does not clearly lend itself to a local or community solution. There is an obvious need for national, regional and local government to take a leadership role and devise a long-term strategy for dealing with air pollution and its effects. The Vietnamese government, for example, could initiate the introduction of unleaded petrol and the gradual phase-out of leaded petrol. Another possible action might include monitoring the pollution emitted by motorcycles and cars, and beginning to implement emission standards. While Ho Chi Minh City ostensibly tests vehicles for emission standards (few vehicles are actually tested), the standard is very low by world standards and the fine for not meeting the standard is quite affordable.

The real dilemma, however, has to do with engagement of Ho Chi Minh City residents. Most residents are apparently not aware of the long-term health crisis stemming from declining air quality, because it is happening slowly and people are adjusting to the ever-worsening air quality (O'Rourke, 2001a). The lack of knowledge regarding pollution and the lack of resident input into potential solutions will hamper the ability of Vietnamese cities to actually mitigate the effects of contaminants. As in many developing country cities, few residents of Ho Chi Minh City understand, for example, the connection between motorcycle use and overall health.

Many people are reluctant to switch to cleaner fuel or use more expensive motorcycle technology primarily because they experience only the cost and not the benefits directly. If the link between health and motorcycles or leaded fuel or dangerous small factories were established and publicized, residents might be more likely to support enforcement efforts and eventually come to demand on them. Community-level initiatives addressing the links among health, productivity, pollution and the economy are virtually non-existent in current Vietnamese environmental policy. Once residents in Vietnam understand and experience the link they are much more willing to organize, act and improve their environments (O'Rourke, 2001b).

O'Rourke's work on a number of industrial sites in Vietnam illustrates community willingness to act to enforce environmental legislation once community groups become convinced of the impact of contaminants on their livelihood or health. One example comes from a textile company located in a newly urbanizing area near Ho Chi Minh City, Dong Nai province (O'Rourke, 2001a). Pollution from the factory, a joint venture with majority Taiwanese ownership, is apparently the cause of respiratory illnesses, rotting roofs and darkened plants in the surrounding close-knit community of Dona Bochang. The most compelling problem for residents, however, is the impact of the factory's harmful emissions on activities related to the community's Catholic open-air church. A number of community events, such as weddings, have been ruined by black soot landing on trays of food or ceremonial clothing and religious items.

The Dona Bochang residents organized to meet with factory representatives, helped to develop newspaper reports of pollution incidence and wrote numerous letters to government representatives. Eventually, the Dong Nai Department of Science, Technology and Environment (DOSTE), the agency nominally responsible for monitoring pollution levels, agreed to take action. Initially, DOSTE limited its role to a planned inspection that bore little fruit, but eventually the factory was convinced through negotiations with DOSTE as well as confrontations with residents (who literally live along the factory's back wall) to build a taller smokestack, change its practice of 'blowing the tubes' from its water heater which was the primary source of the black soot and install an air filtration system to capture the pollution. These three steps have improved local air quality according to O'Rourke's conversations with Dong Nai DOSTE officials and factory representatives.

O'Rourke (2001a) notes that the state's role in this solution was complicated because the Dong Nai People's Committee owns a small share in the factory so that the community's efforts to improve their local environment directly conflicted with the economic interests of members of the committee who, not surprisingly, also control the Dong Nai DOSTE. Yet the community was able to achieve at least some of its goals by using the media and relying on national-level agencies such as the National Environment Agency. The community was really responsible for initiating the process and for maintaining significant pressure on the local and national governments. Yet it was only resident direct knowledge and perception about the impact of air pollution on their quality of life that motivated their concerns. Without understanding that unhealthy air quality was negatively affecting their community in a tangible way and deciding that the factory could afford to take steps to prevent this from occurring, community residents would most likely have done nothing. Lack of knowledge regarding health, and lack of technology to improve air quality, unfortunately hamper most people in Vietnam in initiating actions that will improve local environmental conditions.

It is undeniable that innovative and motivated community groups can do a great deal to facilitate individual responses to environmental laws and required behaviours, and have major impacts on local and perhaps regional abatement of air pollution. However, while O'Rourke (2001b) argues that community action may be sufficient to actually enforce environmental legislation and perhaps replace government action, the regional scale of air pollution creates substantial challenges to solely relying on community mobilization for effective solutions. Instead, we argue that community involvement is probably best seen as one key component in the successful implementation of improved air quality. In addition, what is also required is public, NGO and private sector cooperation in addressing varying sources and types of air pollution. Without the public resources devoted to ensuring the quality of what is perhaps the ultimate public good, air, there is little likelihood that a community can enact (or perhaps more importantly, sustain) sufficient change on its own to protect the air of Ho Chi Minh City. Similarly, the monitoring technology and the enforcement power represented by the law is not currently within the capacity of most local community organizations, and it is problematic to give policing powers to existing and organized community groups, who may act in fragmented and isolationist ways.

Policy Implications

In essence the three brief examples summarized above illustrate several typical environmental management approaches and ongoing problems faced by developing country cities and their urban poor. In each case, Jakarta, Manila and Ho Chi Minh City, the environmental management system was comprised of overlapping and underfunded bureaucracies with major shortcomings in terms of human resources and enforcement capacity. In each case, it would appear that more effective solutions would require the participation of communities so that service delivery could be improved. Unfortunately, the few instances when community participation and input is allowed or even supported represent a small drop in the ocean of environmental pollution created within the city. Most efforts to manage the environment continue to emphasize centralized, impersonal and very inefficient state-centric environmental management solutions that call for residents to stop polluting but do little to facilitate their understanding or their cooperation.

In Jakarta, the primary environmental policy used to address the unequal and inadequate access to potable water in many areas of the city was to increase the availability of standpipes and, in a relatively creative move, to allow homeowners with their own municipal water connection to sell directly to neighbours without a water connection. While this policy helped reduce prices slightly, it is clear that most of Jakarta's water problems remain unresolved. As in many, if not most, developing country cities, access to water is driven by income and the people with the lowest incomes generally have the worst access, consume the least water and pay the most for that water compared to other urban residents. The bureaucracy primarily responsible for providing treated water to urban residents does not have the resources to expand the network to the extent required by rapid urbanization. Therefore, although the vast majority of Jakarta residents are willing and apparently able to afford to pay for adequate supplies of municipal water, that option is not readily available and this is in spite of the attempt on the part of the public entity to involve the private sector.

Solid waste management in Manila has also reached disastrous levels. As in many cities, the revenues or user charges collected for solid waste collection and disposal do not begin to cover the cost of this service. The local, regional and national governments seem incapable of designing and implementing a system that regularly collects most solid residential and commercial waste from city streets and pavements. Consequently, much of the waste remains uncollected, where it pollutes the soil and groundwater as well as contributing to the presence of disease vectors. Most of the existing dumps or landfill sites have closed as they become full and there is an ever-growing reluctance on the part of urban and rural residents to allow new dumps to open within their geographic boundaries. While solid waste disposal in Manila is clearly a critical issue, the local and central governments have not been able to devise and fund workable strategies. The reliance on a task force to develop guidelines and make recommendations has not solved the problem caused by years of inadequate solid waste management. The involvement of the private sector in future landfill construction and management is widely believed by the populace to be part of a corrupt process and the cost of these services is unlikely to be affordable to the majority of Manila's residents.

Finally, Ho Chi Minh City residents have been exposed to ever-worsening levels of air pollutants. While the Vietnamese government has promulgated laws and regulations that nominally protect the environment, the actual resources required to do this are basically lacking. The agencies responsible for environmental protection have overlapping responsibilities, are staffed by inadequately trained bureaucrats and lack the technology to enforce many of the regulations. In addition, the unique blend of capitalistic socialism that Vietnam is attempting to implement results in inherently contradictory situations whereby the government officials responsible for ultimately enforcing environmental protection legislation may legally have an economic interest in entities that are in violation of the law. While Vietnam, as most developing countries, has an interest in protecting the environment, the pressures placed on the government to facilitate economic development are great and often drive the actual agenda. Vietnamese residents are rapidly motorizing, and the externalities generated by the decision to drive a motorcycle or ship goods by lorry are not borne by the polluter. The laws and regulations are currently only important in spirit and not in practice.

The Indonesian, Filipino and Vietnamese cases are illustrative in that they bring to the fore the variety of environmental management policies implemented or used by governments to manage rapid environmental deterioration. Interestingly, in each case, there are examples of approaches to environmental management that are local and community-focused in orientation and that seem to address a number of the ongoing environmental management problems faced by urban governments. In the case of Jakarta, inclusion of household preferences and community input into the design, cost and construction of water supply projects seems to represent an alternative approach to a centralized remote and inaccessible water system. Similarly, a local woman's group in Manila working with and for recyclers has been able to modify the nature of the local and informal scavenger business sector to the extent that it has become a safer and more effective means to collect waste in dense urban areas. Finally, as noted by O'Rourke (2001a) in Ho Chi Minh City, an organized, informed and activist community was able to pressure a local factory into meeting environmental regulations (with the assistance of the local bureaucracy). It would seem that community participation and civic engagement are key components of more effective environmental management policies and strategies.

In the following chapters we explore the nature and form of civic participation in a particular city that suffers many, if not all, of the environmental management problems faced in Jakarta, Manila and Ho Chi Minh City. The city of Bangkok, Thailand, is famous throughout the globe for many positive but also negative attributes, not the least of which are extremely high levels of air, water and soil pollution. While the government of Thailand and the city government (the Bangkok Metropolitan Authority) are generally regarded as *laissez-faire* and are clearly preoccupied with encouraging rapid economic growth, it is also true that the residents of Bangkok have awakened to the dangers and costs of a polluted environment. The potential conflict posed by environmental deterioration and growing evidence of civic participation in response to lackadaisical protection of the environment provides a fascinating case study.

Chapter 3

Environmental Management Policies in Thailand

Environmental management in Thailand is directed and led by several levels of government as well as implemented through the cooperation of individuals and households who comprise the neighbourhoods where environmental issues manifest themselves. This chapter addresses those state agencies and bureaucracies directly responsible for environmental management in Thailand at the municipal, regional and national levels. To provide a context for the more specific discussion of environmental management, the first part of this chapter briefly describes the geography of Bangkok. The following section deals primarily with the formal structure of the institutions in place that oversee the design and implementation of environmental policy in Thailand. As in most aspects of its civil bureaucracy, the Thai state follows a pattern of much formal authority with little capacity for flexibility or even for actually translating policy into timely and effective environmental improvements, particularly in rapidly changing urban settings. The third part of this chapter describes how these formal structures control or shape the attempts of local urban communities to influence Thai environmental landscapes. We argue that bureaucratic institutions – that is, the apparatus of the Thai state – while they purport to seek or increase the role of public participation particularly in the environmental sphere, in fact also work to (un)consciously discourage or impede participation. This argument helps to explain the general failure of most community groups in the Bangkok Metropolitan Region (BMR) to achieve notable improvements in their environmental quality of life. The chapter concludes with a broader discussion of the ways in which these state forms and functions define environmental policy strategies, but also lend themselves to significant intervention by para-state groups and institutions (such as the military and international donors) and non-state groups and organizations (such as private capital interests and non-governmental organizations).[1]

---

[1] We return more specifically to the issue of non-state intervention in environmental policy, specifically by non-governmental organizations, in Chapter 5.

The Geography of Bangkok

The Bangkok Metropolitan Region (hereafter referred to simply as 'Bangkok') is Thailand's capital city as well as one of the most primary cities in the world. Until 1997, it was also the centre of an almost uniquely successful metropolitan region in terms of economic growth. The city experienced rapid population growth in the past decade with an estimated increase in population from 5.7 million in 1980 to 10.8 million in 1990. Bangkok is a remarkable city that offers residents and visitors alike an extraordinary range of urban experiences.

Thailand is a kingdom, with the current king seen as a revered constitutional, but influential, monarch who lives in a modern palace near the historical centre of the city. The country has never been colonized by a European power and perhaps for that reason Thai culture is a truly unique in Southeast Asia. The country has been ruled, in the main, by several elite families for at least seventy years and despite claims to democratic processes, the country is subject to frequent *coups d'etat* where not much changes in terms of political influence except the last names of the people holding Cabinet positions.

Most Thais are Buddhists, and while the physical landscape in the city of Bangkok appears quite modern, the internal social life of the Thai people remains strongly tied to traditional socio-cultural norms that are very different from modern industrialized relationships typical of the West. Two hundred years ago, when Bangkok first rose to national prominence, the city was composed of primarily teak houses and shops, with the main means of transportation constituted by a system of canals. As the city has grown and modernized, it has become a unique blend of traditional and modern Thai cultures, having very positive and very negative dimensions with respect to community environmental conditions. On the positive side, for example, the city is home to a series of beautiful historic temples (*wats*) and is situated on both sides of a major river (the Chao Phraya) that flows into the Gulf of Thailand some 20 miles south of the city (Figure 3.1). And for shoppers and tourists, Bangkok's nightlife and traditional and modern shopping areas are world renowned.

On the negative side, as in many other rapidly developing urban centres throughout Southeast Asia (as discussed in Chapters 1 and 2), Bangkok has experienced extreme water and air pollution, traffic congestion and solid waste disposal challenges. Rapid urban development and modernization have had high costs. The transportation congestion and air pollution problems in Bangkok are considered dangerous to public health as well as costly in terms of economic impacts. The land on which Bangkok is located is gradually sinking, partly due to the expansion of groundwells, and the city wrestles with severe floods during the annual rainy season. Construction of modern

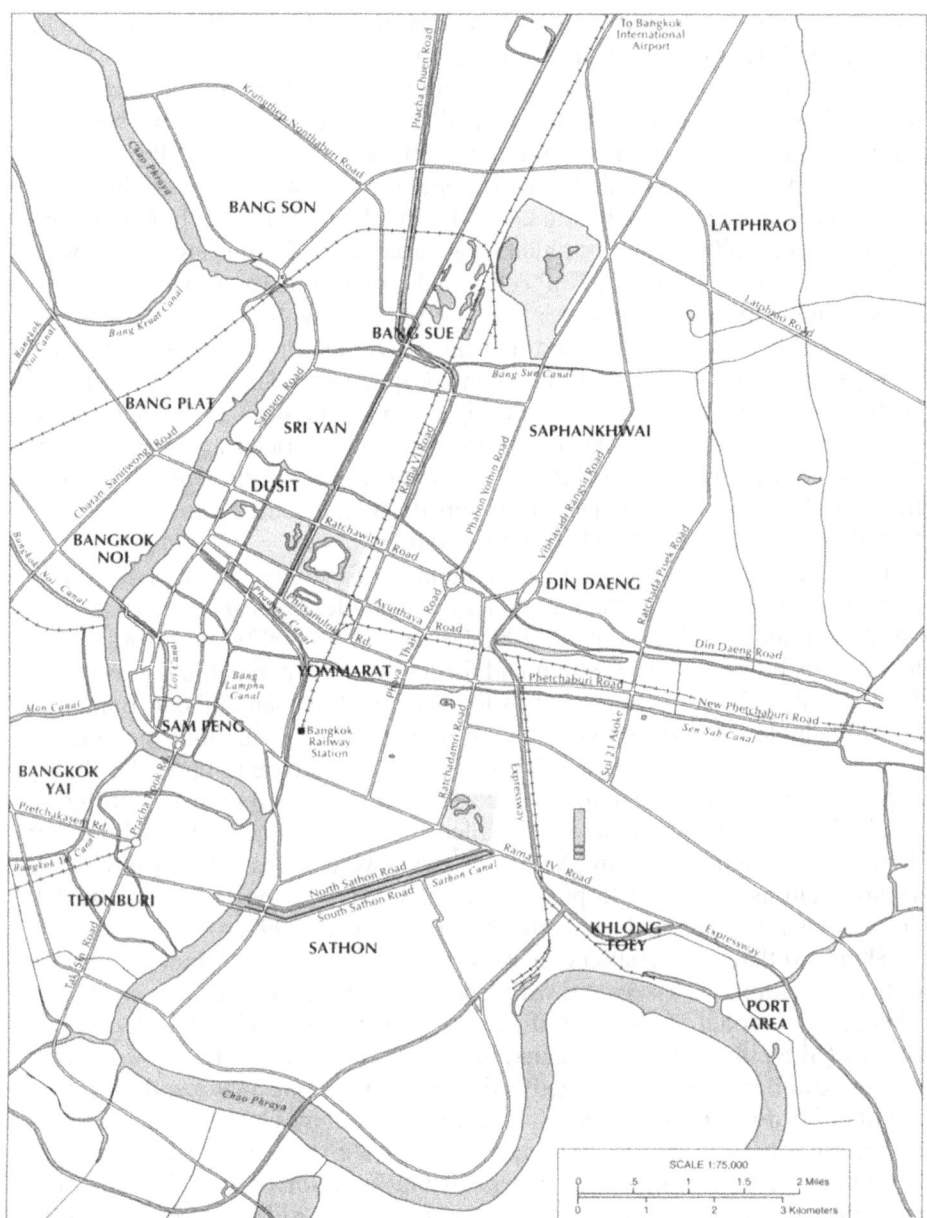

Figure 3.1 Map of Bangkok, Thailand
Source: Ulack and Pauer, 1989

buildings, huge roads and highways, and even an elevated light rail system have created what some consider a modern urban nightmare.

To comprehend the character and magnitude of the environmental problems in Bangkok, it is important to understand the nature of the region's housing, infrastructure, and economic conditions.[2] Physical expansion of the city has paralleled and even exceeded that of population growth, particularly in the central core and the suburban areas of Bangkok. Approximately 630 square kilometres (396,000 *rai*) of rural land was urbanized within Bangkok between 1974 and 1988 (Dowall, 1992). This rapid expansion was largely managed through the formal private-sector housing industry with an apparently lesser role played by informal land and housing providers. Private developers have managed to build a great deal of housing despite rising land and construction costs through modification of the design, types and sizes of housing units. Generally, even the least expensive dwelling units, priced at approximately 250,000 *baht* (about US$10,000) and affordable to a proportion of low-income households, come with connections to piped water, electricity and telephones, and offer decent access to public transportation.[3]

Low-income housing for the lowest quintile of income earners, however, is generally not available in the region. As an example, the least expensive units on the market, which cost about 200,000 *baht* (US$8,000), are affordable to households between the twentieth and fortieth income percentile. However, in 1989 less than 3 per cent of the units for sale cost less than 200,000 Baht, that is, affordable to households below the twentieth income percentile. Consequently, most low-income households in Bangkok end up in the slowly expanding and suburbanizing squatter settlements of the region. According to Setchell (1992), who completed a detailed survey of 968 slum households for Thailand's National Housing Authority, there were at least 1,400 physically separate slums in the metropolitan area that were home to about 300,000 households (approximately 1.7 million persons) as of 1992.

Slum dwellers and squatters are also at a disadvantage in Bangkok because of the vulnerabilities associated with lacking tenure security. Data collected by Thailand's National Statistical Office (NSO), as part of their Socio-Economic Survey (SES) in 1991–92, indicated that between 30 per cent and 60 per cent of slum dwellers in medium-sized and large slums were squatters, that is, they admitted to dwelling rent free on land for which they have no title. Without tenure security, households are less likely to invest in housing. In addition, lack of tenure security can hamper access to the kinds of basic public services typically provided by the public sector to urban residents. More recently, however, landowners have had to contend with the potential for opposition and

[2] The information summarized in this section is culled from recent research as well as a survey conducted in February–March 1994 of over 500 households in Bangkok (Daniere and Takahashi, 1997, 1999a).
[3] Historically, the Thai *baht* was pegged to the US dollar at the exchange rate of 25 *baht* to US$1. Current well-reported problems within Thailand's financial markets have caused this rate to fluctuate significantly higher since the collapse of the *baht* in 1997.

even legal strategies initiated by squatters that may prevent landowners from ejecting squatters from their property; increasingly, sales of private land (on which squatters currently live) must incorporate the potential for lawsuits or other legal efforts to prevent eviction.[4]

It is not surprising that Setchell also found in 1992 that approximately 20 per cent of Bangkok residents lived in settlements with inadequate waste and sanitation facilities, contaminated water and erratic and unsafe supplies of electricity. Data on the Bangkok region, from both Thailand's National Statistical Office and the Setchell surveys, suggested that slum households had access to few of the amenities typically available in formal housing. Even more disconcerting were the types of access problems faced by Bangkok households living in less-established slums, for example, family groups that lived on construction sites or under bridges, without regular access to services. They constituted a section of the population that various public, private or non-governmental agencies found almost impossible to assist.

Thailand
Source: Carrie Mitchell

The inadequate and crowded housing conditions in Bangkok represent the result of rapid, uncontrolled expansion as well as the relatively low incomes of the urban poor. According to international estimates, in 1990 per capita GNP in Thailand was about US$1,570. Per capita annual income in 1990 was probably around the order of US$1,200. Income is always difficult to estimate from surveys, but our estimates of per capita annual income among the urban poor in Bangkok in 1994 were significantly lower, about US$655. Additional information confirms that the socio-economic status of squatters and slum dwellers in Bangkok is quite low. For example, slum households surveyed in 1994 claimed to have an average monthly income of about 10,562 *baht* (US$400) which seems likely given that most respondents (over 90 per cent) had not completed primary school (Daniere and Takahashi, 1999a).

Few observers would deny that Bangkok presents both remarkable opportunities and potential disasters. The city is clearly one of the most exciting places to work and live in the world at the same time as, due to a complex mix of economic, cultural and political realities, there are a number of issues that make it a very problematic place to live. While affordability and traffic congestion seem to be two of the overriding concerns of the middle class, the environmental deterioration of the urban landscape negatively impacts upon the quality of life of everyone, most particularly the urban poor. The rest of this chapter outlines the role played by the Thai state to manage Bangkok's environment.

[4] We thank Dr. Sora Park Tanjasiri for sharing this observation from her recent interactions with Thai private landowners.

The Role of the Thai State

The environmental policies of the Thai state provide the institutional context within which Thai residents develop their environmental perceptions and practices. The Thai state's ability to address environmental issues is partially dependent on the available and effective use of information, on the organization and scope of state authority, and on the varying interest groups involved in political activities. These can be organized into three interrelated components: (1) state organization and interests; (2) organization and interests of different socio-economic groups; and (3) the supportive and conflicting relationships between state and social actors (Skocpol, 1985). As in many parts of the world, the Thai state response to environmental concerns has often been to import solutions and policies from the West without considering the unique geographic characteristics or technical capacity of the Thai kingdom, necessarily addressing existing cultural values, or developing the appropriate incentives and mechanisms to ensure their successful implementation. We describe here the form and function of the Thai state to provide a context for a subsequent institutional analysis of Thai environmental management.[5]

The Nature of the Thai State

The nature of the Thai state is in large part defined by the patrimonial character of Thailand. As in many Southeast Asian nations, within traditional patrimonial Thai society, the king and his agents or representatives historically comprised the top level of the ruling structure. Kings in Thailand granted officials both rank and land but these honours were not accompanied by hereditary rights. Although individual farmers or peasants were not tied to specific plots of land, they were required to work for the king or his local representatives. This series of relationships implied that while there was no permanent aristocracy, there was obligatory labour particularly among the peasant class. Even in modern times, following the overthrow of the absolute monarchy in 1932, Thai society has remained patrimonial in essence. As one would expect, stemming from these historical and cultural practices, well-being and status have continued to be dependent in large degree on the strength of ties with superiors, that is, those embedded in the elite and ruling classes.

Many scholars, such as Mulder (1985), Mizuno (1976) and Klausner (1983), note that Thai communities exhibit a relative lack of reciprocal obligations and an absence of regularly constituted groups. In Thailand, as elsewhere

[5] This section draws on earlier descriptions of the Thai environmental management system written by the authors. Specifically we wish to acknowledge that portions of this chapter draw directly from *Environment and Planning C* 15, Daniere and Takahashi, 'Environmental Policy in Thailand:Values, Attitudes, and Behavior among the Slum Dwellers of Bangkok'. pp. 305–27 (1997), with permission from Pion Limited, London.

in many Asian societies, relationships are notable for the widespread pattern of clientage coupled with the formation of short-term and rather narrowly based social organizations. Typically, these groups are dependent on the leadership of strong individuals (Wurfel and Burton, 1996, p. 35). According to Unger (1998, p. 34), however, Thais are unusual in that they are more apt to exhibit diffidence regarding organized activities and be particularly anxious to avoid reciprocal obligations than villagers in other Southeast Asian countries such as Malaysia and Vietnam.

Historically, the relative absence of horizontal linkages among equals in Thailand did not prevent or inhibit the creation of vertical bonds between individuals and communities. Vertical relationships were omnipresent between patron and client in all aspects of Thai life. All recent evidence suggests that, even today, patronage is very important in almost all Thai social relations. Clientage networks are prevalent among business people, government officials, provincial leaders and traditional Thai elites. We should note, however, that these linkages are subject to frequent changes in composition as individuals are able to shift between networks with some ease. This generally implies that there is an even greater dependence on leaders than in many other places with similar patterns of social relations (Kemp, 1984, p. 59).

The patrimonial character of Thai society explains much of the structure and functioning of the Thai state. The Thai state is often described in two interrelated ways: (1) as a bureaucratic polity, because of the dominant role the bureaucracy plays in policy design and implementation; and (2) as a praetorian political system, because of the role that the military plays in administering bureaucratic activities.[6] The Thai state as bureaucratic polity is well delineated because, as many researchers have shown, most of the governing responsibility resides within the centralized bureaucracy. This centralized bureaucracy dominates both private and public activities, and is composed primarily of the ruling class and the military elite of Thailand. Thailand is, in fact, one of the few democratic countries in the world where the king continues to play a very important role in both the cultural and political spheres. The rapid Westernization and industrialization in Thailand since the Second World War occurred at the initiative of the country's political leadership and bureaucracy. The end purpose of the modernization, however, has not been to transform the traditional system but to preserve and strengthen it. In many ways, traditional relationships and behaviour have not been affected by Thailand's industrialization, and the elites have sustained the traditional system of power relations. The

---

[6] Riggs (1966) was the first researcher to categorize Thailand as a bureaucratic polity followed quickly by Huntington (1972) who emphasized the praetorian nature of the Thai state. This abbreviated discussion of Thai politics relies on more recent texts by Hewison et al. (1993), Ruland (1992), Bunbongkarn (1987) and Jumbala (1987). These authors, as well as a number of other researchers (for example, Girling, 1981; Keyes, 1987; Xuto, 1987), do significantly more justice to the complexities of Thailand's political economy than we can hope to achieve in this summation.

emphasis on preserving traditional power relationships and control are pervasive throughout the upper ranks of the most influential members of the bureaucracy and the military. The most outstanding evidence of how this political regime operates is the regular occurrence of relatively non-violent coups (one on average every three years since 1955) accompanied by little, if any, change in who actually governs the country. Politics in Thailand consequently tend to be dominated by bureaucratic institutions that are insulated from the public at large and, generally, not accountable for their actions. In addition, the duplication of agencies, policies and responsibilities, and the hierarchical nature of Thai society effectively inhibit public criticism of political officials.[7]

As the second characterization of the Thai state (the praetorian political institution) suggests, military influence in governance has been substantial for many decades. The rise of military regimes in Thailand is partially due to the inability of civilian political groups to govern satisfactorily. The fragmentation and internal conflict inherent within past civilian regimes have made military intervention common even during periods when civilians had ample opportunity to exercise political power. Such fragmentation and internal conflict is probably related to the lack of horizontal social ties within general Thai society. That is, civilian regimes might require more horizontal cooperation and collaboration to maintain influence over the Thai state bureaucracy even with relatively influential vertical social relationships between supporters and patrons. The lack of horizontal social ties among vertically connected social groups means that, in general, there is little public resistance to military coups; there have been at least three instances of military intervention since 1960 where public opinion has forced the king of Thailand to intervene directly in the outcome.

Public Administration in Thailand

Effective public administration in Thailand is hampered by the disjuncture between planning and implementation. Such insufficient harmonization is primarily due to the absence of institutions with the capacity to bargain or coordinate either within or among state agencies and other interest groups. In the 1960s, Jacobs was one of the first authors to comment on the weaknesses of the Thai state apparatus. He argued that in spite of the public sector's apparent efforts to assist the Thai industrial sector, such efforts were essentially unsuccessful because they were not guided by a basic rational conceptual framework. Consequently, public officials often attempted to implement conflicting and irrational strategies (Jacobs, 1971, pp. 122–3). As in many Southeast Asian nations at the time, the Thai state was genuinely interested in

[7] For a significantly more detailed discussion of the Thai bureaucracy see Samudavanija (1987).

planned and consistent development. The desire for development, however, was not coupled with a willingness to modify the essential patrimonial and institutionally fragmented nature of Thai economy and society (ibid., p. 137).

In terms of the Thai state apparatus, various authors advance additional explanations for its administrative ineffectiveness. Historically, the Thai bureaucracy essentially achieved independence from political oversight and control when the bureaucrats played a key role in the overthrow of the traditional monarchy in 1932. The bureaucracy has been relatively free of the control of political leaders since that time (Riggs, 1966, pp. 105–7). After 1932, in fact, Neher (1999) suggests that the void created by the absence of political supervision was filled increasingly by the well-known aspects of patrimonial societies, such as manipulation, social networks and allegiance, rather than by merit or administrative competence. Scott (1985) and others describe what they call the 'feudalization' of the Thai public administration since 1932 as a result of Thai leaders giving bureaucratic positions and offices to reward political allies.

These factors together create an environment that permits the state apparatus, or public administration, to function ineffectively and remain unresponsive to the public at large. Interestingly, some writers have argued that such seeming weakness in state structure may actually serve to strengthen state influence over private firms and civil society. Dobbin and Sutton (1998) argued that in the context of the US state, for example, when a state issues ambiguous mandates to organizations, changes rules frequently in response to protracted political negotiations and litigation, and enforces its rules in a fragmented and indecisive way, this state may appear weak and internally inconsistent. However, these 'weak' state characteristics also produce a peculiar kind of state strength. Firstly, when the state leaves the terms of legal compliance unclear, organizations and agencies will need to commit substantial resources to ensure that they are compliant with the possible judicial and legal interpretation of vague regulatory statutes. Secondly, because the state signals uncertainty about the legitimacy of its own authority, organizations and agencies may have to devote resources to developing rationales for those compliance strategies that they are continually obligated to implement (Dobbin and Sutton, 1998, p. 1).

While Unger (1998) and others make the point that such a decentralized system allowed a variety of different organized interest groups to have some input into decision-making, what it did not facilitate, however, was the access of the public at large, and in particular marginalized communities, to processes of public policy formulation or implementation. Recent political events strongly suggest that the emerging urban middle class is beginning to develop linkages across groups and is gaining access to the public administration bureaucracy. As in the past, however, political parties in Thailand continue to experience widespread factionalism. The relative lack of linkages among existing organizations representing varied interests acts to minimize the appeal of these political parties to labourers and peasants. As such, any citizen participation that does occur tends to centre on policy implementation rather than policy design and development.

Any actual impact of community groups and organizations on policy design and implementation appears to be minimal. There are ad hoc examples where public opinion and action has altered prevailing environmental policy. Public outcries regarding household waste disposal in the city of Chiang Mai, for example, resulted in a decision not to use a 'degraded' forest area as a new dump site. The city had had a problem of where to dispose of its municipal waste for several years and had decided, with the support of the provincial forestry office, to use the forest area as the new dump. Protests by local environmental groups resulted in the selection of a more suitable and acceptable location with fewer negative environmental impacts. However, most individuals in Chiang Mai and other municipal areas of Thailand remain ignorant of how environmental policy is made and how policy affects their health and community environmental quality. Household waste may remain uncollected for weeks or even months in some areas, and may be disposed of in unsafe dumps rather than monitored sanitary landfills. Such actions still take place despite a variety of laws that purport to ensure that landfills in Thailand must meet strict environmental standards.[8]

Another example of lack of capacity to enforce laws that have a major impact on environmental and public health is the recent leakage of extremely radioactive cobalt-60 stolen from a radiotherapy machine stored in a car park in suburban Bangkok. The Kamol Sukosol Electric Company was charged for violating the 1961 Atomic Energy for Peace Act because it did not notify authorities of the transfer of these machines from a warehouse to a zinc-walled car park and because it failed to store the cobalt-60 in a safe place. In addition, the company neglected to inform authorities that the lead containers containing the material had disappeared several months prior to the leakage.

In the meantime, however, at least ten individuals were exposed to potentially fatal doses of the substance, because the lead container was cut open in a nearby scrapyard by an individual who could not read the English warnings on the container. News reports regarding the leakage caused many in the immediate neighbourhood to visit local hospitals to be tested for exposure. Unfortunately, while the laws exist in Thailand to prevent and sanction companies that engage in such environmentally dangerous actions, the public role in the implementation or application of these laws is limited at best. Although a number of innocent, and not surprisingly, low-income residents may die from this lack of information, knowledge and enforcement, the public outcry is limited and often led by the media (Pongpao and Ngamkham, 2000).

---

[8] We return to the issue of community participation in environmental management in Bangkok in Chapter 4.

A few non-governmental organizations, including the Alternative Energy Project for Sustainability, GreenNet and Foundation for Consumers, have criticized the Office of Atomic Energy for Peace for poorly managing the radioactive waste leak, failing to evacuate residents and failing to contain a contaminated area. These organizations have contended that an agency that cannot handle a small leak such as this one should not be put in the position of managing a nuclear reactor currently being planned for the Nakhon Nayok province north of Bangkok. These arguments appear to be falling on deaf ears, however, as the government recently announced that there would be no delay in the construction of the reactor (Wancharoen, 2000).

The ability of the masses to influence or control environmental policy appears to be very limited even when there are organized groups attempting to represent at least a segment of public opinion. The capacity of low-income marginalized populations to affect environmental management policies even in the city of Bangkok is further constrained by lack of knowledge regarding environment and health issues as well as the barriers to participation reinforced by the traditional patrimonial structure of Thai society. The capacity of low-income populations to participate in governance of their own communities is further limited by the Thai state apparatus, which has historically denied a decision-making role to residents having little economic clout. How does the Thai state bureaucracy cope with the worsening problems of air, water and solid waste pollution that have challenged economic and social well-being particularly over the past decade? This chapter now turns to a description of the state bureaucracy to outline and investigate the structure of the Thai state apparatus that deals with environmental issues.

Environmental Bureaucracy[9]

The centralized institutions that characterize the Thai bureaucracy are similarly common to the water supply and sanitation sector, often resulting in the ineffective use of resources and personnel. A complex bureaucracy and hazy lines of responsibility create a situation of overlapping agency functions, territorialism, and 'top-down' design and implementation strategies. Broadly, the existing systems suffer from a range of problems including limited funds, technical deficiencies, ineffective regulation, land scarcity, traditional household environmental management practices that resist change, a lack of engagement by the public around these issues and, as always, a lack of political will and commitment (UNESCAP, 1995).

The management and implementation of water supply in Bangkok is primarily the responsibility of two agencies. The Metropolitan Waterworks

---

[9] The following section draws heavily on material previously published in Daniere and Takahashi, 1999b, 'Public Policy and Human Dignity in Thailand: Environmental policies and human values in Bangkok'. *Policy Sciences 32*: 247–268 with the express permission of Kluwer Academic Publishers.

Authority (MWA) is responsible for delivering piped water to all residents and businesses inside the area governed by the Bangkok Metropolitan Authority (BMA). Both the MWA and the BMA report to the Ministry of the Interior (MOI), one of the most powerful ministries in the Thai political system. The Provincial Waterworks Authority (PWA), which is also under the direct supervision of the MOI, is charged with supplying water to all other areas of the country. In general, the PWA has offices and projects in all of the major cities and towns of Thailand outside of Bangkok (Figure 3.2, Table 3.1). In addition to the two public and strictly regulated waterworks authorities, the MOI also oversees the Public Works Department (PWD) which is responsible for the development, operation and maintenance of sewerage and drainage systems.[10]

There are, furthermore, a myriad of centralized authorities in the Thai bureaucracy that play at least some role in the development and implementation of water supply and sanitation services (Table 3.1). These agencies clearly have a variety of overlapping responsibilities, and often report to completely different ministries. Interestingly, until December of 1994, no agency was directly responsible for sewerage and waste management at the provincial and local levels, but since then the BMA has identified priority zones for sewage connections to treatment plants to be constructed during the next decade. These zones appear to have been chosen on the basis of density and pollution levels. Although two of the treatment plants are more or less operating, it is still unclear what the BMA's policies are regarding the connections of households with informal land tenure arrangements (a significant issue for low-income urban dwellers).

Policy and the Environment

In terms of legislation, the Thai state has a number of policies and areas of action that purport to protect the environment. The Thai state, as represented by the bureaucracy, clearly pays at least lip-service to the environment: forest loss, land erosion and degradation, pollution, congestion in Bangkok, national parks and endangered species are all common coin in the currency of government policy and planning rhetoric. The Seventh National Economic and Social Development Plan (1992–96), for example, focuses many of its recommendations on the area broadly recognized as the environment and advocates implementing the Polluter Pays Principle across many types of programme initiatives (NESDB, 1991).

In 1992, the government created a new Ministry for Science, Technology and Environment with much more nominal power than the National Environment Board it replaced. At about the same time, the government passed the extremely detailed Enhancement and Conservation of National Environmental

---

[10] Krongkaew (1990) describes a 'triangular' relationship among municipalities, the PWD, and the PWA which are empowered to work together to deliver piped water and sanitation services to individual cities if requested by the municipality.

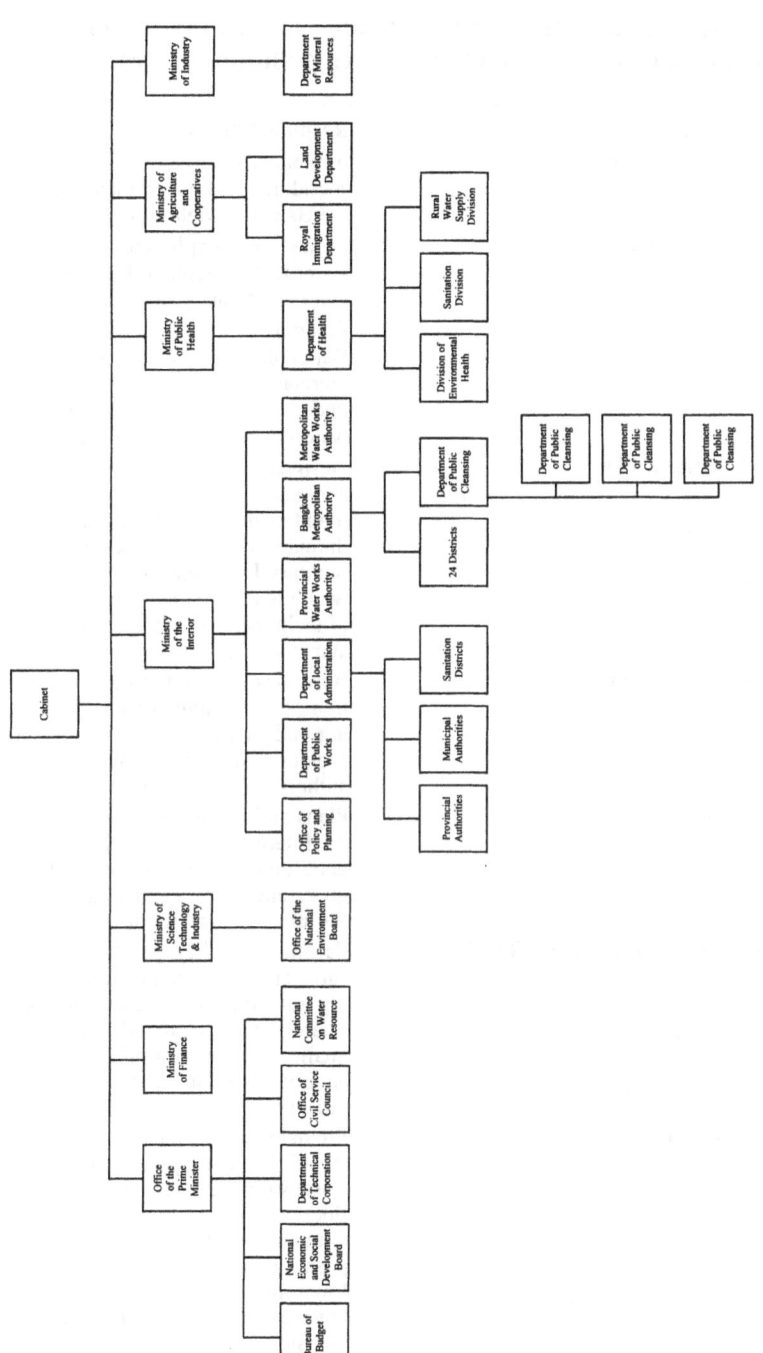

Figure 3.2 Principal Water Supply, Sanitation and Waste Management Agencies

Sources: UNESCAP, 1991; Unkulvasapaul and Seidel, 1991.

Table 3.1 Agencies and Responsibilities (water supply, sanitation, wastewater management and water pollution control)

Agency or committee	Responsibilities
Bangkok Metropolitan Authority	Implements activities of corresponding local government; acts in the Bangkok Metropolitan Region
Metropolitan Waterworks Authority	Production, supply, and sale of water to metropolitan Bangkok and neighbouring areas
Provincial Waterworks Authority	Piped water supplies outside Bangkok area; undertakes commercial operations in urban areas; undertakes construction of small water-supply systems in rural areas
Department of Public Works	Designs, supervises construction, and advises on operation and maintenance of sewerage and drainage systems; houses the Sanitary Engineering Division which is responsible for provision of technical assistance on wastewater treatment
Department of Local Administration	Activities involved in the operation of local municipalities including budget and finance; houses the provincial authorities, municipal authorities and sanitary districts
Office of Policy and Planning	Plans and formulates Ministry of the Interior policy; coordinates plans and programmes of local governments, especially natural resources management
Office of the National Environment Board	Formulates policy, plans and measures on the environment including guidelines for implementation; establishes effluent and water quality standards; administers the environmental impact assessment process
Department of Health	Provision of technical services for installation and operation of wastewater treatment facilities, flood protection and storm water drainage, and hazardous waste disposal; houses the Division of Environmental Health, Sanitation Division, and Rural Water Supply Division

Royal Irrigation Department	Provides water for irrigation; constructs dams
Land Development Department	Land development, especially for agricultural purposes
Department of Mineral Resources	Manages groundwater
Office of the Permanent Secretary	Controls the provincial policies; appoints provincial governors
Bureau of Budget	Allocation of government budget
National Economic and Social Development Board	Development of policies and preparation of national five-year development plan
Ministry of the Interioir	Implements water, sewage and wastewater management policies and programmes; oversees the living conditions of Thai people; responsible for all activities at the provincial and local levels
Ministry of Agriculture and Cooperatives	Rural land use and development, national resource protection and management
Ministry of Industry	Promotes industrial development and controls industrial pollution; provides water supply from groundwater, and controls groundwater extraction
Ministry of Science, Technology and Energy	Implementing agency for conducting scientific research and development
Ministry of Public Health	General public health, control of wastes, and nuisance abatement through public health sections within municipal governments
Department of Technical and Economic Cooperation	Coordination of external assistance, both technical and economic; important for obtaining external assistance in terms of financial support, training, technology transfer and equipment
Office of Civil Service Council	Allocation of number of staff and their salary level to government agencies at all levels
National Committee on Water Resources	Formulation of water resource development

Quality Act B.E. 2535 in 1992. The Act adopts an integrated and multidisciplinary perspective in its approach to environmental problems. For example, under the terms of the Act, the role of non-governmental organizations in environmental management is described and endorsed. Furthermore, the Act gives the Ministry the right to allocate clean-up and environmental protection funds through a new Environmental Fund and decentralizes control of the environment to the provincial level.

While the Act increases the potential for funding environmental management policies and enforcement, this chapter has noted earlier the substantial extent to which individuals and entire ministries ignore plans and regulations; this obviously limits the potential effectiveness of environmental management policies. Rigg (1995) argues that in addition to the fractured working relationship among ministries, there is a great deal of corruption and cronyism that are probably synonymous with the patron-client networks that abound in Thai society. In particular, as stated by Trébuil (1995), 'successive [Thai] governments have treated environmental problems like political disputes', which undoubtedly prevents government agencies from acting in the public's best interest. An example of this process is provided by Bangkok's transportation system and traffic congestion; both of these issues are severe and widely acknowledged as among the worst across the globe. Yet, no government agency and no combination of ministries have been capable of putting the interests of Bangkok's residents ahead of the private economic interests of powerful corporations and individuals (Daniere, 1995).

Thus, the primary problem with environmental policy-making in Thailand is not necessarily one of policy formulation, although there is obviously a lack of public participation and access at this stage of the policy process. Rather, the major impediment to responsive governance that acts to protect and enhance the environmental quality of Thailand's communities is the lack of capacity to implement policy and/or enforce regulations. In the last decade, the Thai state apparatus, in response to international criticism as well as domestic protests, has given great lip-service to the inclusion of the public and different communities in all phases of policy design and implementation. The actual achievement of this goal, however, remains elusive.

Bureaucratic Response to Environmental Problems

The institutional structure of the agencies charged with providing water and sanitation services to the Bangkok Metropolitan Region (BMR) is partially responsible for the poor provision of these services in many parts of the metropolitan area. Water in the region is supplied to consumers through a combination of groundwater and surface-based systems. In outlying areas, water supplies are drawn from deep wells; in central Bangkok, the transition to a piped surface water-sourced system is nearly complete. Unfortunately, the MWA has not been able to expand piped water services at a pace anywhere near that of urbanization. Consequently, many households in outlying areas must either forego public water connections for an average period of seven years or rely on developers who sink their own deep wells and contribute to the region's substantial land subsidence problems (Daniere, 1996).

Water quality in the Chao Phraya River has also suffered greatly during Bangkok's process of rapid urbanization and industrialization. In 1990, the dissolved oxygen concentration was, on average, 0.5 to 1.0 milligrams per litre, well below the standard recommended for household consumption and industrial utilization purposes (2.0 milligrams per litre). In 1994, the BMA began operating the Si Phraya wastewater treatment plant, the first of its kind, which currently handles approximately 6 per cent of all wastewater generated in the BMR. However, many challenges remain to sustain the operation of this plan and to expand wastewater treatment throughout the city. One large obstacle centres on public response to user fees needed to fund the operation of this and future plants. Public response is unlikely to be positive given the lack of incorporation of residents in the design and implementation process, and the relative public inexperience with having to pay for public infrastructure services. The authority faced public protest when it proposed to charge residents for these wastewater treatment services in 1995 (Sriswaskraisorn and Bamroong-ua, 1994).

Solid waste collection in Bangkok is much less formal and complex than the water and sewerage systems in that collection within low-income settlements is primarily organized by the residents.[11] Communities can request that the BMA locate a dumpster close to their settlement and pick up the household waste deposited there once or twice a week. The BMA is quite responsive to resident requests but essentially responds in a reactive manner. In addition, very few resources are expended to limit the extent of vermin and pests within the city, and communities are not provided with incentives to clean up their communal areas and pathways.

The disposal of solid waste is complicated by the lack of sanitary landfills and incinerators. The city currently relies on unlined dumps to dispose of its refuse, causing health risks due to potential groundwater contamination, vermin and concentration of disease. The dumps, however, also provide economic opportunities for low-income communities through scavenging. Thus, improving the environmental conditions of these dumps and preventing scavengers from working there might enhance the health and quality of life of low-income settlements living nearby at the same time as it would make earning a living for other low-income dwellers more difficult.

Finally, environmental quality issues overall are the responsibility of the Ministry for Science, Technology and Environment which oversees the

---

[11] Most of the discussion regarding solid waste disposal services comes from Luanratana (1994).

Pollution Control Board (PCB). The institutional structure of the National Environmental Board (NEB) was reconfigured in the mid-1990s presumably to enhance the effectiveness of environmental regulations. Although the restructuring of the board was prompted by a growing awareness on the part of Thai policy-makers of the economic consequences associated with deteriorating environmental conditions (especially in terms of tourism and business opportunities), the agency did not seem to have been given greater enforcement powers. The organization thus had the *de jure* authority to set environmental standards but lacked the resources to actually implement and enforce the improvements in air and water quality. Furthermore, the motivation behind the restructuring of NEB seemed to be directly related to the economic costs of increased pollution on business rather than the rapidly deteriorating and increasingly dangerous living situations faced by the urban poor.[12]

The Thai State and Environmental Governance

This chapter has sought to describe the institutional form and bureaucratic policies and practices of the Thai state as they relate to environmental management. What should be clear from this description is that the tangled structure of the Thai state bureaucracy creates situations where unclear responsibility and overlapping substantive and spatial jurisdictions define daily governmental practice in terms of environmental management. The top-down nature of the Thai state, the emphasis on technical solutions to environmental degradation, and a lack of resources to implement and enforce environmental policies contribute to a disjuncture between formalized state policy and effective environmental management at the community, municipal, provincial and national scales.

Scholarly treatments of environmental management more generally have also tended to emphasize issues largely concerned with developing technical solutions to address environmental problems. Bryant and Wilson (1998) in their recent review and critique of this literature argued that such scholarship remains largely focused on: (1) top-down approaches that discount grassroots, local, and non-scientific knowledge and interests; (2) state-centric formulations of environmental management that overlook the role of non-state actors, groups and organizations; and (3) top-down, scientific and traditional methods, actors and power relations that leave out the structural and institutional relations defining problems, stakeholders and appropriate solutions. There have been efforts to alter this state-centric and technical focus. Andrews (1999) points out, for example, that the environmental management literature has recently addressed more geographic, political, social and economic dimen-

---

[12] Panswad (1987) and personal interview with Nisakorn Kositratna, the Director of the Water Quality Management Division within the Ministry of Science and Technology, in January 1995.

sions in terms of coping with environmental problems, allowing policy-makers and researchers to analyse in a more comprehensive manner the challenges and opportunities of managing environmental issues in a context of rapid economic, cultural and political change.

Thai state actors have realized that dealing with environmental issues is an increasingly vital component to sustaining economic growth in Thailand. But accomplishing this goal in a political context of patronage, corruption and cronyism is difficult at best. In addition, existing and proposed environmental management policies and practices are significantly directed by the military elite and non-state actors and organizations, adding to the confusion and ineffectiveness already embedded within the bureaucratic structure. Existing state-centred views on government are limited in their applicability in Thailand for several reasons. As already discussed, the Thai state bureaucracy emerged as a direct consequence of historical, cultural and social norm structures that sustained the moral influence of the monarchy, and the hierarchical power relations defined by the military and Thailand's social elite. That is, because of its particular 'moral ethos', the Thai state's formation, form and functions will be a direct consequence of Thailand's historical, political-economic, social and geographic conditions (following Corrigan and Sayer, 1985). Moreover, the Thai state has been increasingly focused on economic growth throughout the 1990s, especially given the economic downturn, requiring a significant degree of interaction with private sector interests both domestic and international to spur and sustain development. Because of these unique issues and the ever-increasing influence by international donors and non-governmental organizations in environmental politics, a conceptual approach to environmental management is necessary that incorporates the probable role that non-state actors (for example, private capital interests but also non-governmental organizations, ad hoc grassroots groups, and individuals) may play in defining environmental management policies and practices. The concept of environmental governance is offered now as a way to clarify and articulate the role and practices of state, para-state, and non-state actors in environmental management in Thailand.

Toward Environmental Governance

Governance perspectives centre on the notion that politics have become 'destatized' (MacLeod, 1999, p. 345). That is, in contrast to existing state-centred or society-centred views on government that focus on the form and function of the state, governance perspectives emphasize the coordination and conflict among interdependent activities and actors, both state and non-state, and their actions and practices in steering policy. Consequently, in studying governance, emphasis is placed on the interrelationship of institutions, stakeholders and actors involved in issue formulation, policy design, implementation and enforcement:

> We conceive of governance as a set of institutions that may comprise varying configurations of actors situated, for example, in the state, the market, and/or civil society. Institutions are composed of sets of rules, which may be formal or informal, codified or implicit. These rules define institutions in terms of constituent actors, the conditions for their inclusion, their reach or authority, the states of the world they can affect (i.e., their jurisdiction), the flow of information, the mechanisms through which decisions are produced and/or overturned, and the distribution of end results (NSF Workshop on Urban Sustainability, 2000, p. 18).

Economic development has been the primary focus of such theorizing in scholarly treatments, especially in Western industrialized nations, where partnerships (public and private), the creation of quasi-public agencies and the privatization of public activities have most clearly illustrated the linkages and networks between the state and markets (for example, Jessop, 1998; also Rhodes, 1997):

> First, governance can refer to any mode of co-ordination of interdependent activities ... Its forms include self-organizing interpersonal networks, negotiated interorganizational co-ordination, and decentred, context-mediated inter-systemic steering. The latter two cases involve self-organized steering of multiple agencies, institutions, and systems which are operationally autonomous from one another yet structurally coupled due to their mutual interdependence. (Jessop, 1998, p. 29).

Governance has also been defined as

> a tendency for the role of the official state machinery in securing state-sponsored socioeconomic projects to diminish, in favour of a variety of partnership arrangements between governmental, paragovernmental, and nongovernmental organizations and the expansion of 'regulated self-regulation' (MacLeod, 1999, p. 352).

Jones (1998) divides the literature on governance into two distinct paradigms, both with the aim of identifying, characterizing and critiquing 'new urban governance structures' (p. 965): (1) regulation theory (for example, analyzing changes in the state given uneven temporal and spatial transitions); and (2) what he calls 'local sociologies of translation' encompassing the three perspectives of regime theory, rational choice theory and policy networks, all of which focus on 'networked negotiation, agenda setting, interorganisational and intersystemic steering' (p. 963). In this second paradigm, Jones defines regime theory (and its parallel work on growth machines) as focused on 'the political economy of place and networked co-operation between the public and private sector within local economic development'; this framework is included under the rubric of New Urban Politics (NUP) (p. 962). Rational choice, as defined by Jones, entails the argument that 'policy making is a game involving a series of trade-offs and decisions based on the location and strategy of other actors' and involves bargaining by powerful actors and groups (p. 963). Policy networks, Jones's third component of his 'local sociologies of translation', and focused primarily on the work by John and Cole, emphasizes 'the way that local networks mediate and structure contrasting institutional arrangements in local economic policy making' (p. 963).

MacLeod and Goodwin (1999) depict the literature on urban governance focusing on the more land-use oriented dimensions of the literature, organizing current scholarship into two groups: urban regime frameworks (for example, 'the tendencies for regimes to emerge out of civic cooperation based on mutual self-interest between governmental and non-governmental actors', p. 701) and growth machines analysis (for example, where coalitions form because of the conflicts between the use and exchange values of land). Growth machines were popularized by John Logan and Harvey Molotch (1987) who argued that the differences between the 'use' and 'exchange' value of land prompted distinct actions by growth machine and antigrowth proponents. Cox and Mair (1988) described such regional growth coalitions as 'relations of local dependence' because in encouraging growth, members of these growth machines must make compromises, which in turn 'contain the seeds of future difficulties for jurisdictional projects' (p. 14; also Cox and Jonas, 1993).

These conceptual arguments incorporate the notion that the state has taken on entrepreneurial aspects that had previously been thought largely confined to the private sector. Especially in the realm of economic development, state actors and bureaucratic institutions have not only partnered with private and other non-state organizations, state agencies themselves have become much more entrepreneurial. Scholarship on such public sector entrepreneurialism has tended to emphasize state functions that link most directly with private sector urban development, for example, local economic growth to improve the aggregate income and private sector growth potentials for a locality or region as a whole (for example, Cox, 1993; Leitner, 1990; Mollenkopf, 1983). In practice, such 'entrepreneurial governance' has been much less focused on than welfare and collective consumption (which serve to enhance the quality of life for the labour force), on the distributions of wealth resulting from public sector investments or on local management of environmental resources (that is, the 'use value' of land; Logan and Molotch, 1987).

Research on political and policy entrepreneurship, in contrast, largely promoted by political scientists and emanating from the public choice and pluralist perspectives, has focused on elected and non-elected bureaucratic officials, arguing that political entrepreneurs behave in parallel ways to private sector entrepreneurs in trying to radically change current policy debates and enhance their political power and influence. From this perspective, political entrepreneurs develop strategies for marketing ideas and policies to others, and create and sustain coalitions and networks to support such ideas (for example, Baumgartner and Jones, 1993; Eisinger, 1988). To fulfil these objectives, political entrepreneurs use 'policy windows', where 'streams of problems, policies and politics' coincide, enabling redefinitions of problems and innovative solutions (Kingdon, 1984; also Lober, 1997). Policy windows require a confluence of opportunities and individuals able to take advantage of these opportunities for innovation to occur.

Given this set of writings as a starting point, the concept of environmental governance would begin from the argument that in seeking problem redefinitions and innovative solutions, the state will tend to seek opportunities to partner in new ways with groups and organizations that were previously overlooked or not included. Furthermore, state agencies might begin to act in more entrepreneurial ways to design and implement environmental management policies, such as privatization, where environmental governance practices might even be centred in non-state arenas and initiated largely by non-state actors and organizations. The role of non-state actors and organizations has not been widely explored by researchers and policy-makers. As Bryant and Wilson (1998) argue in their review of the environmental management literature, non-state actors and organizations, although they 'appear regularly in conventional accounts', are largely seen as 'objects of state attention – to be consulted or manipulated by state environmental managers in keeping with official policies' (p. 325). However, such non-state policy entrepreneurs would seem to play a significant role in the formation and potential institutionalization of environmental regulation at multiple scales of governance. In the Western industrialized world, the significant role of non-state policy entrepreneurs is relatively clear in the growth machine literature, as pro-growth advocates who represent the social and political privilege of the growth machine (Elkin, 1987; Mollenkopf, 1983).

In the US, anecdotal evidence has indicated that non-state individuals and organizations not promoting private sector development interests have played an increasing role in environmental problem definition and policy creation more generally (Bartholomew, 1999). To accommodate this expanding role of non-state individuals and organizations in the developing world, especially those not representing private capital interests, an environmental governance framework should articulate the form and practices of networks of international donor organizations and national or provincial environmental activist organizations, but also small grassroots local groups and individuals, in environmental regulation debate, design, implementation and enforcement.

Predominant conceptual approaches to governance, especially in urban planning and geography, tend to focus on economic governance in the Western industrialized world, and 'examine the practices and conflicts which emerge from such institutions [engaging in economic regulation] and the ways in which particular expectations and behavior ... are stabilized and regularized through conventions, compromise, and, ultimately, the exercise of power' (Jessop, 1997, cited in MacLeod and Goodwin, 1999, p. 706). To clarify environmental governance in Thailand and the developing world more generally, especially at the local level, analysis must focus on the networks of collaboration, cooperation and even conflict, that enable the design, implementation and enforcement of national, provincial and local regulation and management of environmental resources.

Such institutional relationships can provide opportunities for policy innovation when collaboration and cooperation define interaction among actors. However, the definition and institutionalization of policy opportunities, and even of collaboration, may also serve to limit the realm of conceivable interaction both in terms of the substantive issues to be addressed and the appropriate actors to be involved (following Cox and Jonas, 1993). To explore the potential of an environmental governance perspective in understanding environmental management in Thailand, we turn now to two forms of non-state action that may both reinforce and resist current institutional forms and practices of centralized bureaucratic environmental management: community action (Chapter 4) and the activities of non-governmental and ad hoc community organizations (Chapter 5).

Chapter 4

Environment and Community Participation

Although greater community participation is often explicitly identified by international donor agencies and many nation states as integral to the successful management of local environmental issues, the values, attitudes and behaviours of local residents regarding environmental issues in different urban areas across the spectrum of the developing world remain poorly understood. In particular, although there have been increasing calls for 'demand-driven planning',[1] relatively little is known about local perceptions of both the benefits and costs of changing environmental quality. The ramifications of this lack of knowledge are quite significant. Existing community practices may serve to exacerbate institutional problems in the provision of water and sanitation services. Most city dwellers in the developing world have limited experience with piped and treated water, flush toilets and modern sewage treatment facilities. They are consequently reluctant to pay for such services and, even though the resulting environmental conditions are quite dangerous and unhealthy, they do not demand these services from an increasingly remote and technocratic government. Thus, many governments in the developing world have been unable to design financially viable water and sanitation systems that meet the social welfare needs of residents while promoting economic development and protecting natural resources.

---

[1] Whittington et al., 2000, p. 297; studies of communities in developing countries have examined issues related to (1) the role of factors influencing the choice of water source where the poor have many feasible alternatives at hand, ranging in price, convenience and quality; and (2) cost recovery issues, emphasizing the question of whether the poor are willing to finance facilities that might not otherwise be built. Studies performed by the international donor organizations financing the expansion of water systems in these countries, such as the World Bank and the US Agency for International Development (USAID), have focused on the latter as they attempt to reform the public infrastructure investment process in these countries. Taken together, these studies argue that (1) if cheap, satisfactory alternatives are available, such as clean well water, the poor will not opt to connect to a piped municipal system even at very low cost, and (2) if clean water is scarce, the poor will be willing to pay extremely high prices for good water, and will be likely do so to private suppliers, perhaps enough to finance an expanded piped system (Briscoe et al., 1990; Whittington et al., 1991).

To address these issues, particularly as they relate to Bangkok's low-income slum dwellers, this chapter is organized into three sections. Firstly, we provide a brief overview of the issues surrounding community participation in developing countries, especially in terms of the scholarly treatment of this issue. Secondly, we analyse the linkages among cultural values, attitudes and perceptions, and behaviour and action among Bangkok's low-income urban dwellers. This analysis is based on a survey of 540 urban dwellers that we administered in 1994. Finally, the chapter concludes with implications for understanding more fully the connections among cultural values, attitudes, individual action and environmental governance.

Understanding Community Attitudes and Participation

Much scholarly work on community attitudes in the developing world in the neoliberal tradition has fallen under the rubric of 'community participation' (Whyte, 1984; but see Bebbington, 2000 for a critical review of this literature). Researchers within this literature argue that knowledge about community attitudes and behaviours can be used in a range of activities from consultation with the community about intended projects to self-sufficient community development (Midgley et al., 1986). There are several themes that emerge from this literature: community participation includes a variety of activities; the ultimate purpose of community participation is to increase the effectiveness of development projects; and community participation is difficult to promote (White, 1981; Kalbermatten et al., 1982; Hogrewe et al., 1993). There is an expanding body of research, especially by Dale Whittington and his colleagues, that measures individual attitudes concerning environmental management issues; however, for the most part, scholarly understanding remains limited of the relationship in the developing world between community attitudes and practices and environmental governance issues (such as the provision of water and sanitation services).[2]

This lack of knowledge and specific information about community perceptions and action has hampered many efforts to mobilize local resources, particularly in places such as Thailand where community participation has historically been very limited.[3] There is limited understanding of the behaviours and attitudes of the majority (that is, low-income households frequently living in the most polluted areas of cities) in terms of what they

[2] Qualitative studies of this issue have focused on the role of the family in sanitation implementation, hygiene practices, knowledge of sanitation issues, social norms regarding defecation and personal health beliefs (for example, Moser, 1993; Yacoob et al., 1992).
[3] Cumper (1984), Bomba (1982) and the World Bank (1998a) suggest that attitudes towards and the use of sanitation comprise the critical factor in the successful provision of this type of public service, rather than the mere physical provision of facilities. Nakata (1987) elaborates on the traditional lack of citizen participation.

believe about the environment, the role they attribute to Thai state bureaucracies in managing environmental issues at multiple levels, and the extent to which they are motivated to pay or work for environmental improvements.[4] Thus, in Thailand, service delivery as well as environmental policies and regulations that depend upon local participation and support are frequently designed and implemented without incorporating the interests of those groups most likely to be called upon to participate.

In terms of environmental governance, community participation may well serve a more insurgent role. Grassroots actions around environmental policies and management have become increasingly visible particularly in the US, where 'environmental justice' has rallied communities of colour and marginalized groups to protest against maldistributions of environmental hazards (for example, Pulido, 1996). Thus, increasing the knowledge and potential participation of low-income residents concerning water, sanitation, and solid waste issues might not only facilitate individual and neighbourhood participation in Thai state-supported activities such as community clean-up and infrastructure construction, but could also catalyze low-income Thai urban dwellers, largely disenfranchised and excluded from the political decision-making process, to initiate local collective action.

Consequently, in the Thai context, an important initial step in understanding the potential for community participation, whether cooperative or insurgent, consists of the characterization of perceptions and behavioural intentions regarding the role of Thai state bureaucracies and local residents in environmental management and governance, especially in Thailand's rapidly urbanizing areas. Research suggests that there are a number of important relationships among values, attitudes and environmental behaviour that need to be clarified if governments are to successfully implement environmental protection measures (for example, Jones, 1996). An examination of the literature reveals that within developing contexts relatively little is known about the environmental behaviour of urban residents, particularly the poorest of these residents, and how environmental behaviour is impacted upon by values, attitudes and socio-demographic characteristics.

One strategy to address this gap is to apply attitude-behaviour models used widely in the Western industrialized world that allow a detailed portrait of attitudes and behaviour among slum households in Bangkok regarding environmental practices, health and community participation. Researchers in Western industrialized contexts have long looked to attitudes and public perception as significant precursors to behaviour and action, especially with respect to environmental management and protection. Attitude-behaviour models, most widely associated with Ajzen and Fishbein's (1980) Theory of Reasoned Action, posit that individual, communal and societal characteristics

---

[4] Whittington et al. (1993) make this point about sanitation systems in a number of developing countries but present very detailed data for residents of Kumasi, Ghana.

frame the potential realm of possible attitude and belief structures toward specific issues and objects, which then in turn affect individual intentions to act regarding those specific issues and objects, resulting in eventual action or non-action. Using this general perspective, scholars and policy-makers primarily in Western industrialized contexts have worked to predict behaviour based on assumptions that specific attitudinal structures will eventually lead to behavioural outcomes. For example, attitudes toward environmental issues or governing institutions may impact upon a household's capacity and willingness to participate in environmental projects or initiatives and thus may have important policy implications in terms of the design and implementation of environmental management and governance. In this chapter, we use the attitude-behaviour framework to examine the linkages among specific cultural values, socio-economic characteristics and environmental practices especially among low-income residents in Bangkok for issues such as participation in community projects or initiatives, water treatment, waste disposal behaviour and health practices.

The attitude-behaviour framework promoted by Ajzen and Fishbein (1980) suggests that in part specific socio-demographic characteristics will significantly impact upon attitudes and behaviour. For low-income Bangkok residents, for example, individuals with higher levels of formal education might be more likely to participate in community clean-up efforts as well as more likely to take measures to protect their health through treatment of drinking water and regular disposal of solid wastes in community bins, because of their enhanced understanding of the connection between contaminated water and individual and household health.

In addition to the linkages between attitudes and action, we also pay particular attention in this chapter to the influence of Thai cultural norm structures on attitudes and behaviour. Community participation in Bangkok in particular is likely to be influenced by traditional Thai values such as individualism (for example, Thai individualism refers to the primacy of the individual and his/her household or family rather than the more widely known Western version of individualism, which tends to focus on the promotion of self-interest), avoidance (for example, an individual is uncomfortable with the idea of confronting problems or issues directly), or a strong support for the patron-client relationship that permeates Thai social relations (for example, an individual is supportive of hierarchical relationships between patron and client). It is reasonable to suppose, for example, that someone who holds more traditional Thai beliefs is less likely to take action to improve any aspect, including the environmental quality, of his or her community. Conversely, someone who is less supportive of traditionalcultural values may well be more inclined to participate in community-wide improvement strategies. Education, at least non-traditional education, may well play a role in changing or influencing cultural values that might discourage participation in actions aimed at improving environmental infrastructure or conditions.

To examine these and other linkages among widely accepted Thai cultural values, individual perceptions, and environmental behaviours and actions, this chapter now presents results from a survey of 540 low-income urban dwellers in Bangkok conducted in 1994 that focused specifically on these linkages.

A Survey of Low-Income Dwellers in Bangkok[5]

We conducted a survey of low-income households in the BMR to gather information that might explain the potential linkages among environmental behaviours and actions, attitudes and perceptions, and cultural values of individual households. An initial version of the household questionnaire was developed with Thai colleagues (Professors Orathai Ard-Am and Anuchat Poungsomlee) from Mahidol University in January 1994. Approximately ten household interviews and open-ended discussions were conducted with respondents in two different slums. The questionnaire was then pre-tested with twenty-five households.

In total, we surveyed 540 low-income urban dwellers in Bangkok during the winter of 1994. The respondents were all residents of slum and squatter settlements and included 232 persons from the inner zone of the BMR, 238 persons from the middle zone and 45 individuals from the region's outer zone. Surveys were administered by advanced students from Mahidol University during February and March of 1994. The enumerators were each given intensive training in the administration of the questionnaire.

The settlements were stratified according to the following criteria: the three municipal zones (inner, middle and outer); established squatter or slum settlements within these zones; the division of settlements into household units. Residential location by zone was important because access to services depends heavily on location. Outer areas, in particular, have less access to water and waste disposal services than other parts of the BMR. Focusing on established areas (that is, settlements that have existed over a substantial period of time) created a sample primarily composed of households having knowledge about environmental conditions and activities in their communities.

The survey was administered to the person in the household responsible for making most of the decisions regarding water and sanitation issues. Consequently and as we expected, a majority (more than 78 per cent) of the sample was composed of women. Most of the women respondents (73 per

[5] Much of the following section comes from material previously published in Daniere and Takahashi, 1999a, 'Environmental Behaviour in Bangkok, Thailand: A Portrait of Attitudes, Values and Behaviour' in *Economic Development and Cultural Change* 47(3): 525–557 with explicit permission from the University of Chicago Press.

cent) were spouses of the head of household although approximately 12 per cent of all women respondents identified themselves as the household head.[6]

The socio-economic status of the sample of squatters and slum dwellers in Bangkok we surveyed was quite low (Table 4.1). In general, as alluded to in Chapter 3, respondents in this sample were somewhat less educated and poorer than average for Bangkok residents but, overall, the socio-economic characteristics of respondents were comparable to other survey results collected from the urban poor. Income is always difficult to estimate from surveys, but our estimates of per capita annual income among the urban poor in our Bangkok sample were significantly lower than international estimates, about US$655. The average household in our sample had a monthly income of about 10,562 baht (US$400) and most respondents (over 90 per cent) had not completed primary school.[7]

Table 4.1 Summary Characteristics of Respondents

Percentage male;female	21.3%; 78.7%
Percentage household heads	32.4%
Average years of schooling	5.81 years
Average age of respondent	37.56 years
Average number of people in household	5.088
Average household earnings per month	10,562 baht (US$420)
Average expenditure on housing per month	762 baht (US $30)
Average health care expenditure per month	515 baht (US $20)
Average monthly transportation expenditures	1250 baht (US $54.00)
Average food expenditures each month	4200 baht (US $168.00)
Average expenditures on water each month	248 baht (US $10)
Average expenditures on monthly solid waste collection	20 baht (US $0.80)

N=540 unless otherwise stated.
Source: Daniere and Takahashi (1999a).

[6] White, et al. (1972) is generally regarded as the source of the seminal discussion of the pri-mary role of women in water-related spheres. Additionally, women have been recognized as having decision-making authority over the household's water and sanitation practices in developing countries in a number of subsequent studies (see, for example, Crane and Daniere, 1996).
[7] As of December 1994, the poverty line in the BMA was approximately 3,000 *baht* per month. This suggests that most households (90 per cent) in the sample do not live in condi-tions of abject poverty. On the other hand, the average household income for the region in 1994 was 18,540 *baht* per month which is obviously much more than the average in our sample households.

To assess the relationship between socio-demographic conditions and attitudes toward environmental problems, besides the socio-economic data just described, we also collected information about the attitudes of respondents towards a host of issues (including the important daily problems in their community, their understanding of the relationship between the environment and physical health, and their attitudes towards governmental problem-solving and responsibility). The survey instrument also included questions regarding the behaviour of individual respondents *vis-a-vis* the environment, health, public services and community participation. Respondents were asked whether they had access to piped water, how they treated their water, how they disposed of their waste, health behaviour, and whether they knew of or had participated in community organizations (particularly in efforts to clean their community or provide it with environmental services). We hypothesized that a variety of factors including household characteristics, value systems and attitudes may all further our understanding of specific actions. Thus, we collected information on socio-demographic conditions, behaviours, beliefs and cultural values, and attitudes from every respondent.

Behaviours: Water and Sanitation, Health, Community Participation

As part of the survey, the individual respondents were queried about their households' behaviours and actions *vis-a-vis* the environment. The queried types of behaviours included whether or not households took precautions regarding their drinking water, disposal practices for wastewater and solid waste, health behaviour and participation in activities organized for the benefit of their community, particularly projects that had an environmental dimension. Table 4.2 presents summary information on the behaviours that are most relevant to environmental action or activity.

Table 4.2 Behaviours

Question	Response	%
Piped water connection in house?	Yes	64.1%
	No	35.9
Sources of water? (includes households who use several different sources of water, usually for different purposes, i.e. washing vs. drinking).	In-house piped	83.3
	Vendor	8.4
	Standpipe	1.2
	Neighbour	1.9
	Well/River/Rain	21.3
	Other	2.6
Boil drinking water? (assuming that household uses piped, rain or well water for drinking)	Yes	45.7
	No	54.3
Use bottled or filtered water for drinking?	Yes	47.5
	No	52.6
Use a BMA bin (rubbish bin) to dispose of your solid waste?	Yes	68.7
	No	31.3
Visited a doctor the last time had a serious illness or an accident?	Yes	67.5
	No	32.5
Visited the doctor many times in the past 2 years?	Yes	71.7
	No	27.3
Children have visited the doctor at least once in the past year?	Yes	64.9
	No	35.1
Have participated in a community group or activity in the last year?	Yes *	22.0
	No	30.0
	Don't Know	48.0

N=540 unless otherwise stated.

* Of the respondents who participated in community groups, 108 individuals participated in community cleaning projects and 69 respondents participated in organizing community-level festivals and celebrations.

Source: Daniere and Takahashi (1999a).

Most of the sampled households in Bangkok (about 83 per cent) had access to private piped water in their homes. As indicated in Table 4.2, a substantial number of households, however, also relied on other methods either to supplement their piped water supply or as a main source of water, including bottled water from a vendor (8 per cent) and water from a private well, the river or rainwater (21 per cent). Most households connected to the piped water system had legal and working meters (64 per cent). On average, the sampled households estimated that they spent about US$10 a month for water including bottled water for drinking that costs, on average, about US$ 0.40 a litre. A slight majority of respondents used piped water for drinking but 47 per cent of these low-income respondents bought bottled water or filtered their piped water before drinking it. Of those respondents who drank piped water, some 46 per cent boiled their water before drinking it.

A majority (75 per cent) of the households surveyed stated that the waste-water from their latrines or water closets flowed directly into open sewers, although independent verification suggested that approximately 50 per cent of all slum communities surveyed had some kind of septic system in place. Additionally, more than 85 per cent of respondents did not have access to individual solid waste collection services; however, most households used a BMA bin or container near their settlement to dispose of solid waste. Individuals pay little or nothing for this service, but even so, approximately 31 per cent of all respondents did not use or have a BMA bin, and dumped their solid waste in open areas within their community.

Respondents were asked a series of questions regarding their health behaviours and beliefs. Important differences in behaviour were noted in terms of how often individuals visited a doctor in the past two years, whether or not respondents went to a doctor the last time they were seriously ill and whether or not they took their children to a doctor on a regular basis. Almost 30 per cent of respondents, for example, had not been to the doctor once in the past two years while 27 per cent of respondents claimed to have visited the doctor many times during the same period. Similarly, 32 per cent of respondents did not visit the doctor despite having a serious illness or accident. Most of those who did go to either a state or private hospital or clinic to see a doctor claimed to have purchased pharmaceutical drugs on their own.[8]

Finally, the data collected from questions centering on community participation indicated that 51 per cent of all respondents were aware of community-based organizations in their neighbourhoods but only 22 per cent actually participated in a group or activity in the last year. The most common types of participation included a community cleaning activity (67 per cent), joining or organizing a savings group (14 per cent), and fund-raising (12 per cent).

Attitudes

The survey instrument included a number of questions regarding the attitudes of respondents towards social and environmental problems in their settlement as well as questions regarding their knowledge of the linkages between environmental degradation and overall health. Furthermore, we incorporated a series of questions that were designed to elicit opinions regarding governmental problem-solving capability and responsibility (Tables 4.3 to 4.7).

---

[8] Antibiotics as well as a wide range of more exotic traditional medicines are generally available without a prescription throughout Thailand.

Table 4.3 Attitude towards Quality of Life

How do you feel that your quality of life has changed over the past year? Has it improved, stayed the same or declined?	
Improved	38.0%
Same	46.9%
Declined	15.2%

Source: Daniere and Takahashi (1999a).

Table 4.4 Attitude towards General Environmental Conditions

How do you feel that the general environmental conditions in your neighbourhood (e.g., air, water, cleanliness) have changed over the past year? Are they better, the same, or worse?	
Better	25.9%
Same	26.7%
Worse	46.9%

Source: Daniere and Takahashi (1999a).

Table 4.5 Attitudes towards Daily Problems II

How much of a problem is each of these issues to you on a daily basis? (solicited responses)

Issues	Not a problem at all (%)	Extremely big problem (%)
Mosquitoes, rats, vermin	3.3	82.4
Odours (bad smells)	30.4	27.6
Rubbish	46.7	25.4
Drugs	41.3	23.1
Traffic	45.9	23.0
Neighbours	85.0	2.4
Clean Water	83.3	3.1
Family and child problems	73.3	3.7
Housing	69.6	12.2
Poor health	60.4	7.4

Source: Daniere and Takahashi (1999a).

Table 4.6 Attitude towards Polluted Water

Do you believe that there is a connection between your illness (gastroenteritis) and the quality of water in your neighbourhood?	
No	54.3%
Yes	11.3%
Don't Know	34.4%

N=540 unless otherwise stated.
Source: Daniere and Takahashi (1999a).

Table 4.7 Attitudes towards Community and Government Action Regarding the Environment

Do you believe that people:		
1. In this community can improve water/sanitation conditions in the community itself?	81.7%	Yes
2. In this community can improve water/sanitation conditions in Bangkok, at large?	66.7%	Yes
3. Who work for the BMA can improve water/sanitation in your community and Bangkok?	70.6%	Yes

N=540 unless otherwise stated.
Source: Daniere and Takahashi (1999a).

Respondents indicated overwhelmingly that one of their major problems was the pervasiveness of mosquitoes, rats and vermin in their environment. In addition to vermin, respondents cited issues related to water (bad water, untreated wastewater), social problems (drugs, evictions, congestion) and other environmental problems (noise, air pollution, dust). When asked specifically about the degree of importance of social and environmental problems, over 80 per cent of the respondents stated that mosquitoes, flies, rats and vermin were a very large problem for their household. This was followed by another set of environmental issues including bad odours, rubbish and noise (although only about 25 per cent of the sample considered these issues to be large problems; Table 4.5). Surprisingly, there has been very little media attention devoted to these topics (compared to issues such as housing and transportation) nor, at least to our knowledge, are there any existing environmental policies addressed specifically at curbing pests and vermin in low-income communities.

We were also interested in the health attitudes of respondents. Our survey indicated that 12 per cent of the total sample had experienced diarrhoea or gastroenteritis during the previous week. Thirty per cent of the total sample of respondents had experienced at least one bout of diarrhoea within the last month. A majority of respondents, however, did not believe that there was a relationship between gastroenteritis and neighbourhood water quality.

Regarding respondent attitudes towards governmental responsibility and the environment, almost 70 per cent of the individuals surveyed stated that the municipal district offices or BMA employees had at least partial responsibility for improving the environmental conditions within low-income communities. Almost the same proportion of respondents also believed that individuals themselves can play a substantial role in improving the general environmental conditions throughout the BMR, while an even greater proportion (80 per cent) of all respondents believed that individuals from a specific community can make an impact on environmental quality within their own settlement or neighbourhood.

Thai Cultural Values

As noted earlier, this survey also collected information regarding cultural values. We recognized at the outset that using a survey to gauge and decipher cultural values was problematic at best. It is most appropriate to explore values and beliefs through qualitative research methods, such as extensive focus-group activities, in-depth interviews and archival analysis. Unfortunately, because of our limitations in time and funding, we settled for a compromise. Respondents were asked a series of questions meant to measure their beliefs in important Thai values that we hypothesized might affect behaviour related to water and sanitation services. The questions themselves were carefully developed, based on extensive review of social and psychological research on Thai cultural values (both from Thai and other scholars), developed in collaboration with Thai social scientists, pre-tested in several different focus groups within individual slums, and modified to improve their precision and clarity before the actual survey implementation began.[9]

Table 4.8 presents measures of the values including avoidance of conflict, extent of individualism, patron-client relationships, fatalism and delayed gratification included in the questionnaire. The value statements in boldprint are those which divided respondents into at least two groups, that is, elicited strong agreement or disagreement among respondents, and divided the respondents into large groups holding different perspectives. These value measures include avoidance of conflict, individualism and the importance of

[9] A detailed explanation of the process involved in selecting the cultural values to be included and their definitions is provided in Daniere and Takahashi (1997). Some excellent sources of general information on Thai cultural values are Kemp (1984) and ten Brummelhuis (1984).

patron-client relationships.[10] Several of these cultural values proved to be significant predictors of behaviour in these survey data, including the patron-client relationship, the idea of avoidance, and individualism. The data suggested, for example, that contrary to popular understanding about Thai society, low-income urban dwellers in our sample did not believe strongly in patron-client relationships. Approximately 70 per cent of the respondents agreed with the statement 'If my patron asks me to do something unpleasant, I should not do it', indicating that this social norm (which is often cited in scholarly literature as a ubiquitous dimension of Thai society) was not widely prevalent among the surveyed individuals.

Table 4.8 Cultural Values among Squatter Residents

Cultural value statements	Value	% Agree*
A. Children should study hard in school to have a better life	Delayed Gratification	75.8
B. Even though I wear a Buddha Image (or a Rama V locket), that will not bring me better luck	Fatalism	40.2
C. **I am not responsible for other people's problems**	**Individualism**	**49.2**
D. **If my patron asks me to do something unpleasant, I should not do it**	**Patron-client**	**70.4**
E. **Dirty klongs should not be taken so seriously**	**Avoidance**	**29.6**
F. People get AIDS because they have bad kharma	Fatalism	35.8
G. **Nowadays, I live on my own and you live on your own**	**Individualism**	**36.4**
H. You should enjoy life every day	Delayed Gratification	40.3
I. If there are people with AIDS in my community, we have a responsibility for taking care of them	Individualism	45.0
J. Rich people deserve more respect than people without money	Patron-client	15.3
K. You should learn to live with your problems	Avoidance	22.4

N=540 unless otherwise stated.

* Agree includes respondents who strongly agreed or agreed with the cultural value statement.

Bold statements divided groups into two main groups of respondents and could be tested for linkages to attitudes and behaviour. Respondents were asked how much they agreed or disagreed with each state and possible responses included strongly agree, agree, neutral, disagree, strongly disagree.

Source: Daniere and Takahashi (1999a).

[10] Respondents did not disagree enough regarding the other value statements to be divided into distinct groups on the basis of their ratings across the Likert scale (with one meaning "strongly disagree" and five meaning "strongly agree"). Stated somewhat differently, the distribution of respondent ratings for the statements associated with the values of fatalism and delayed gratification were more evenly distributed across categories (for example "you should enjoy life every day" measuring delayed gratification) or concentrated in particular categories (for example "children should study hard in school to have a better life" measuring delayed gratification).

A greater proportion of respondents supported the value of avoidance, which is related to the desire or promotion of 'smooth' relationships in Thai culture. Traditionally, one goes out of his or her way in Thai society to avoid conflict and controversy in deference to elites with the expectation that troublesome issues will either resolve themselves or be dealt with by the appropriate patron or elite group. Two statements in the questionnaire measured the value of avoidance. One statement, 'You should learn to live with your problems', expressed the importance of avoiding even the potential for conflict. An individual who is willing to 'live with his/her problems' is more likely to avoid the potential conflicts associated with taking action to solve individual, community or societal problems.

The second statement designed to measure the value of avoidance had an environmental orientation (for example, 'Dirty *klongs* should not be taken so seriously'). An individual who agrees with this statement is someone who would like to avoid any actual or potential conflict or criticism around actions to deal with the severe problem of water contamination in the city's *klongs*. As an example, any project that seeks to clean or restore the canals of Bangkok requires both the description and deterrence of polluting behaviour, perhaps through censure or punitive action, as well as the organization and implementation of clean-up efforts, many of which would necessarily rely on community or public sector supported participation. Given that the level of contamination of the canal system in Bangkok is a widely publicized and extremely visible problem, individuals who assert that the *klong* issue is unimportant are clearly trying to avoid confrontation and/or the necessity for community action. Taken together the two statements imply that a significant proportion (approximately 25 per cent) of the respondents support the value of avoidance as opposed to that of dealing directly with problems as they become evident.

Another value that proved significant was individualism. Thai individualism is somewhat different from the notion of individualism that is prevalent in North America in that Thais interpret individualism to mean the absence of obligation on the part of individuals to social institutions (as opposed to specific individuals or patrons). The value statements used to measure individualism included one that suggests that individuals are not responsible for other people's problems as well as a statement that alludes to the social divisions among individuals ('Nowadays, I live on my own and you live on your own'). Although large segments of respondents agreed with these value statements, a substantial number of respondents strongly disagreed with the value statements measuring individualism (11 per cent and 21 per cent, respectively).

Examining the Linkages among Cultural Values, Attitudes and Behaviour

To ascertain the linkages among the important cultural values, attitudes and perceptions, and behaviour and action pertaining to environmental management among low-income slum dwellers in Bangkok, our analysis proceeds in the following two steps. The first analytical step relies on means tests to delineate significant differences in socio-demographic characteristics between those respondents who practice a certain environmental behaviour and those respondents who do not. We explore how socio-demographic variables, cultural value assessments and attitudes interact and how they predict behaviour. In the second step we analyse more closely the quantitative contribution of socio-demographic characteristics, attitudes, and cultural values to the specific behaviours by means of multinomial logistic regressions. We believe that the results contribute to an improved understanding of environmental policy problems in Bangkok, specifically as they are conceptualized by the urban poor. Perhaps the most important result of our analysis concerns the underlying relationships between explanatory variables and household "participation in community projects" as presented in Table 4.12.

Table 4.9 presents information regarding the significant differences between respondents who exhibit an environmental behaviour and those respondents who do not exhibit the behaviour. Boiling drinking water, for example, represents a relatively simple way for individuals to protect themselves from waterborne illnesses that are quite common in Thailand. Although the municipal water supply in Bangkok is ostentatiously treated, the public generally suspects it of being less than safe to drink directly from the tap. Thus, we ran a simple F-test to determine how, in terms of socio-demographic variables, respondents who relied on piped water for drinking and boiled their water before drinking differed from those who did not. As indicated in Table 4.9, individuals who stated that they did boil their water before drinking were typically more educated and had higher incomes than respondents who stated that they drank piped water without boiling it first.

Table 4.9 Behaviour and Significant Socio-demographic Differences

Environmental Behaviour	Significant socio-demographic variables	Mean (no)	Mean (yes)	F-test	sig.
Metered Water Connection (n=540)	Age in years	34.49	38.76	14.53	.000
	Years of education	5.21	6.15	9.39	.002
	Income (baht p/month)	4759	7859	17.43	.000
	Tenure in Slum (years)	10.69	16.96	29.01	.000
Boil Water Before Drinking (n=293)	Years of education	4.97	6.11	8.23	.004
	Income (baht p/month)	5175	6697	6.53	.011
Use Bottle or Filtered-Water for Drinking (n=527)	Age (in years)	38.18	35.89	4.34	.038
	Years of education	5.33	6.31	10.96	.001
	Income (baht p/month)	5670	7778	8.83	.003
	Tenure in Slum (years)	15.5	13.1	4.67	.031
Dispose of solid waste in a bin or rubbish bin (n=540)	Tenure in slum (years)	13.2	15.4	3.17	.076
Visited a doctor the last time were seriously ill (n=536)	Age (in years)	35.2	38.2	6.75	.010
	Income (baht p/month)	5481	7334	5.73	.017
	Tenure in slum (years)	12.7	15.8	6.42	.012
Have visited a doctor many times in the past two years (n=534)	Age (in years)	36.30	39.59	7.29	.007
	Gender (men=0, women=1)	0.74	0.89	13.92	.000
	Tenure in slum (years)	13.9	17.1	5.80	.020
Have taken child to a doctor at least once in the past year (n=373)	Age (in years)	40.91	36.72	10.90	.001
	Years of education	4.70	6.03	15.36	.000
Participated in a community project during the past year (n=281)	Tenure in slum (years)	14.48	17.08	3.17	.076

Source: Daniere and Takahashi (1999a).

Households who typically purchased bottled water for drinking or used a filter to treat their drinking water differed in several additional ways from those households who did not take additional measures to protect themselves from polluted or contaminated water. In general those who purchased bottled water or filtered their drinking water were younger, more educated, had higher incomes and had lived for a shorter period of time in their slum community than households who did not treat their water or buy bottled water. Respondents who engaged in the third sanitation-related behaviour, that is, getting rid of solid waste in a bin rather than simply dumping wastes in any convenient location, proved to be longer-term residents than those households who did not use a bin.

In terms of health-related behaviours, respondents who were more likely to visit doctors when they were ill tended to be male, significantly older, had higher incomes and had lived for a longer period of time in the slum community than those who were less likely to visit a doctor. Respondents who took their children regularly to the doctor, however, were significantly more likely to be younger and more educated than those who did not.

The final behaviour presented in Table 4.9 relates to community participation. Community participation is one important way in which individuals can act to achieve specific improvements in the quality of their environment. Those respondents who were aware of the existence of community organizations in their slum or settlement were asked whether or not they had participated in any community group activities during the previous year. Those respondents who said that they had participated were found to have lived in the community longer than those who did not participate.

Taken together, these basic relationships between behaviour and socio-demographic characteristics suggested that proactive behaviour in terms of water and sanitation issues, health and community participation were driven by several factors. Education level, for example, appeared to be strongly related to individuals treating their water and to protecting children's health. In addition, longer length of residence in a particular community made it less likely that people acted to protect themselves from contaminated water at the same time as it made them more likely to use proper means of solid waste disposal and more likely to participate in community events. Finally, gender did not appear to be strongly related to a specific behaviour except that women were less likely to seek medical attention when they were ill than men. These findings were quite consistent with what one might hypothesize given the existing literature on education, environmental action and gender differences in health practices in developing nations.

Several logistic multiple regression models were estimated to examine the relationship among household behaviour, socio-demographic characteristics and attitudes and values of the respondents. The dependent variable in each case was the respondent's reported behaviour or, in other words, whether or not a respondent claimed to have engaged in a particular behaviour or not. In the case of whether or not respondents boiled their drinking water (BOIL), for example, the variable had a value of 1 if the respondents reported boiling their piped water before drinking it and a value of 0 if the respondents reported not boiling their water before drinking. The independent variables (Table 4.10) used to explain respondents' behaviour included respondent socio-demographic characteristics (for example, gender, education), household characteristics (for example, household expenditures), measures of respondent attitudes (for example, whether the respondent believes that general environmental conditions have improved or not during the past year), and measures of respondent cultural values (for example, whether a respondent strongly believes in patron-client relationships).

Table 4.10 Description of Variables

Variable Name	Variable Description
Sociodemographic Characteristics	
GENDER	1 = if female
	0 = if respondent was male
SCHOOL	Number of years of education completed
SPE_ALL	Household's monthly expenditures (in baht)
TENURE	Number of years household has lived in slum/ squatter settlement
ZONE	1 = if respondent lives outside of inner core
	0 = if respondent lives in the inner core
Attitude variables	
AIRPOLLU	1 = smoky/polluted air is a major problem
	0 = if not a major problem
EARNINGS	1 = earning a living is a major problem
	0 = if not a major problem
FAM_CHI	1 = family and children are a major problem
	0 = if not a major problem
GARBAGE	1 = if garbage is a major problem
	0 = if not a major problem
GENENV	1 = if general environmental conditions are worse this year than last year
	0 = if not worse
HEALTH	1 = if poor health is a major problem
	0 = if not a major problem
HELPBKK	1 = if people in community can improve water/ sanitation conditions in Bangkok
	0 = if people cannot
HELPCOM	1 = if people in community can improve water/ sanitation conditions within the slum
	0 = if people cannot
HOUSING	1 = if housing is a major problem
	0 = if not a major problem
NEIGHBOUR	1 = if neighbours are a major problem
	0 = if not a major problem
Value variables	
VALUE_C	1 = strong belief in individualism
	0 = not a strong belief in individualism
VALUE_D	1 = support of patron–client relationship
	0 = no support of patron–client relationship
VALUE_E	1 = strong belief in avoidance
	0 = limited/no belief in avoidance
VALUE_H	1 = support of immediate gratification
	0 = support for delayed gratification
VALUE_I	1 = strong belief in individualism
	0 = not a strong belief in individualism
VALUE_K	1 = strong belief in individualism
	0 = not a strong belief in individualism

Source: Daniere and Takahashi (1999a).

Tables 4.11 and 4.12 present the results for the eight model specifications that best predicted the eight specific behaviours. The models were derived in a similar fashion in that all the available socio-demographic, attitude and cultural value variables were originally included in the models and then eliminated on the basis of how much they contributed to the overall ability of the model to predict behaviour.[11] As a result, each behaviour was predicted by a slightly different group of independent variables although there were some important similarities between them. We should note that the overall fit of all eight models was excellent with highly significant chi-square values. In addition, an important result illuminated by the logistic regression models was that attitude and cultural value variables were often highly significant predictors of environmental behaviour.

Table 4.11 Multinomial Logistic Results for Water and Sanitation Behaviours*

Independent variables	Meter	Treat**	Bottle	Getrid
	B Wald Exp (B)	B Wald Exp (B)	B Wald Exp (B)	B Wald Exp (B)
GENDER	–	0.1134 (12.1) 1.120	–	0.4023 (2.83) 1.495
SCHOOL	**0.1070 (7.78) 1.113**	**0.1548 (8.21) 1.167**	**0.0828 (8.84) 1.086**	**0.0742 (6.20) 1.177**
SPE_ALL	**0.0001 (7.21) 1.000**	0.0001 (11.0) 1.0001	**5.16(e-5) (6.05) 1.000**	–
TENURE	**0.0318 (10.65) 1.032**	-0.0154 (3.93) 0.985	**-0.0225 (7.96) 0.978**	–
ZONE	–	-0.3892 (6.24) 0.678	-0.2835 (3.76) 0.753	**-0.4639 (9.03) 0.629**
FAM_CHI	**-1.3009 (4.53) 0.272**	–	–	–
RUBBISH	–	–	–	**-0.5298 (5.73) 0.589**
HEALTH	–	–	**-0.7509 (3.62) 0.472**	–
HELPCOM	–	–	0.4864 (3.27) 1.626	–
HOUSING	–	-0.8894 (3.31) 0.411	–	**-0.6626 (4.96) 0.516**
NEIGHBOUR	–	–	–	-1.0876 (2.85) 0.337
VALUE_D	–	**0.8701 (4.44) 2.387**	–	–
VALUE_H	–	–	–	**0.9643 (20.36) 2.623**
Constant	**-1.0230 (9.78)**	**-0.9985 (7.44)**	-0.5117 (1.62)	0.7163 (3.46)
% predicted correctly	69.14	74.37	67.45	68.91

Source: Daniere and Takahashi (1999a).

* **Bold** typeface indicates significance at the 5% level.

** Treat includes all households who boil or filter their drinking water or who report that they generally purchased bottled drinking water.

[11] The independent variables were all tested for correlation with one another prior to estimating the logistic regressions. As a result of these tests, for example, the age of the respondent was not included in any of the models since age proved to be strongly correlated with both years of education and length of tenure in the slum community.

Table 4.12 Multinomial Logistic Results for Health and Participation Behaviours*

Independent variables	Illness	Visitdoc	Childoc	Particip
	B Wald Exp (B)	B Wald Exp (B)	B Wald Exp (B)	B Wald Exp (B)
GENDER	–		**0.6079 (4.15) 1.170**	–
SCHOOL	–	**-0.0747 (3.70)** 0.928	**0.1567 (13.42)** 1.837	–
SPE_ALL	**3.61(e-5) (3.02) 1.000**	–	–	–
TENURE	0.0177 (5.12) 1.018	–	–	0.0200 (3.51) 1.020
AIRPOLLU	–	**0.8410 (5.91) 2.319**	–	–
EARNINGS	–	**0.7917 (5.94) 2.207**	–	0.7173 (3.19) 2.049
GENENV	**0.4121 (4.21) 1.510**	**0.7130 (7.80) 2.040**	–	–
HEALTH	–	**1.0977 (5.93) 3.000**	–	–
HELPBKK	–	–	**0.4922 (3.56) 1.636**	–
ODOURS	**0.4205 (3.24) 1.523**	–	–	–
VALUE_C	–	–	**-0.9703 (5.87)** 0.379	–
VALUE_D	–	–	–	**0.7381 (4.51) 2.092**
VALUE_E	**-0.3599 (2.99) 0.698**	–	–	0.5341 (3.38) 1.706
VALUE_I	–	–	–	**-0.9090 (9.21) 0.403**
VALUE_K	0.4306 (4.64) 1.538	**-0.5710 (4.63) 0.565**	–	-
Constant	**-0.0947 (0.17)**	**-0.9273 (9.62)**	**-1.0172 (6.10)**	-1.1958 (0.60)
% predicted correctly	68.45	69.37	72.96	65.45

Source: Daniere and Takahashi (1999a)

* **Bold** typeface indicates significance at the 5% level.

The wealth of information presented in Tables 4.11 and 4.12 is fairly straightforward in its interpretation. In general, the left-hand side of a logistic regression is 'logit' or 'log odds ratio'. For example, in the case of the logistic regression model that predicts whether or not the household has a water meter (METER), the model is:

$$\text{Logit (METER)} = \text{Constant} + B1*SCHOOL + B2*SPE\_ALL + B3*TENURE + B4*FAM\_CHI.$$

The 'B' associated with each independent variable is the portion of the log odds ratio due to that variable. In the logistic model that predicts whether or not a household has a water meter, the log odds ratio associated with the number of years of education completed by the respondent (SCHOOL), for example, is 0.1070. The 'Wald' associated with each independent variable refers to the Wald statistic, which is in this case similar to a t-statistic, and provides an estimate of the significance of the variable. In the case of the education variable (SCHOOL), the 'Wald' is equal to a 7.78 value, which is likely to occur by chance less than 1 per cent of the time alone. The 'B' and the 'Wald' value for the years of education variable (SCHOOL) are highlighted to indicate that it is a significant explanatory variable (likely to occur by chance less than 5 per cent of the time).

The final value presented for each independent variable is the exponential of B (the log odds ratio) or the simple odds ratio associated with a one-unit change in the independent variable in terms of the dependent variable. In the case of years of formal education, for example, the 'Exp(B)' value of 1.113 means that a respondent with five years of education is 1.113 times more likely than a respondent with only four years of education to have a metered piped water connection. Additionally, a respondent with six years of education is $(1.113)^2$ or 1.239 times more likely to have a metered water connection while a respondent with seven years of education is $(1.113)^3$ or 1.379 times more likely to have a metered water connection than the original respondent with only four years of education.

In the case of a dummy variable, such as the gender of the respondent (GENDER) in the logistic regression model that predicts whether or not a household disposes of solid waste in a bin or dumpster (GETRID), the odds ratio has a somewhat different meaning. In this case, the 'Exp(B)' value of 1.495 simply means that women are 1.495 or 1.5 times as likely as men to dispose of solid waste in an environmentally appropriate manner.

As noted above, each of the logistic regression models for each of the eight environmental behaviours differs slightly in terms of the significant independent variables. As an example, the likelihood of a household treating its drinking water in some way before drinking it is predicted by years of education (more likely to treat water given more education), income (more likely to treat with higher income), number of years of residence in the slum (less likely to treat water given longer periods of time in the community), zone of residence (less likely to treat water if living outside of the inner core area), how the respondent perceives housing (less likely to treat water if they believe that housing is a major problem), and belief in patron-client relationship (less likely to treat water if they are the norm of patron client relations). In this case, all the variables included in the model are significant at the 95 per cent level.

Similarly, the likelihood of a respondent regularly buying bottled or filtered drinking water (a subset of those who treat their drinking water) is also best predicted by years of education (more likely to buy bottled water as years of education increase), income (more likely to buy bottled water as average monthly expenditures increase), tenure (less likely to buy bottled water as years in the slum increase), the zone (less likely to buy or filter water if living outside the inner core), attitudes towards health (less likely to buy bottled water given the belief that health is a major problem) and attitudes toward community action (more likely to buy bottled water given the belief that individuals can improve overall environmental conditions in the local slum). All but the last of these independent variables (HEALTHCOM) are significant at the 95 per cent level of confidence.

Treating or taking proactive action against drinking contaminated water has been shown in the literature to be strongly associated with improved health (for example, Caincross, 1990). This behaviour appears to be best predicted by socio-demographic variables such as years of education, income and location although attitudinal variables also play a limited role in explaining this type of action. The two models taken together suggest that education is a key explanatory and policy variable in enhancing the likelihood of environmentally protective behaviour. In addition, respondents who believe that individuals can make a difference in the quality of the slum environment also appear to be more likely to protect themselves via improving the quality of their water supply. However and somewhat counter-intuitively, respondents appear less likely to buy bottled water if they believe that health is a major problem facing their household. Such households are likely to be poorer and less educated than those who do buy bottled water and may not, in fact, be as aware of the causal link between health and uncontaminated water as more educated households or may not be able to afford to treat their water in any way except boiling it before drinking. As a result, while they suffer from poor health, these respondents may not have the resources (either in terms of finances or knowledge) to act to improve the quality of the water they drink.

The most important behaviour for the purposes of this chapter is whether or not a respondent participates in community activities. Interestingly, the length of time (TENURE) one lives in a particular community is the only socio-demographic variable for predicting participation in community activities; the B associated with length of tenure is 1.02 with a Wald of 3.51 – a result that will occur by chance approximately 6 per cent of the time. Furthermore, the logistic regression implies that an additional year of tenure increases the likelihood of participation by 1.02 or that someone who has lived in a community for five years is $(1.02)^5 = 1.1041$ times more likely to participate in community activities than someone who has only lived there for only one. Intuitively, this finding makes a great deal of sense: that is, someone who has lived in a community for a long period of time is understandably more willing to invest effort in improving the quality of that community's environment.

Only slightly less statistically significant than length of tenure is whether or not a respondent considers overall earnings to be a problem for the household (EARNINGS). This variable has a B of 0.72 (Wald 3.19), which could occur by chance approximately 7 per cent of the time. The Exp(B) value in this case implies that respondents who stated that income or earnings are a major problem are 2.049 times (more than twice) as likely to participate as those who did not state that income or earnings are a major problem. This is very interesting because it appears that the respondents who are least satisfied with their income are more apt to participate in community endeavours. Duesenberry's (1967) relative-income hypothesis suggests that these are likely to be the poorest people in the community although the income variable (SPE_ALL) did not prove significant in the model when it was included.[12] However, it may well be that more educated and wealthy households actually have the time and willingness to participate and that these households, because of higher aspirations or expectations, are also more likely to state their dissatisfaction with their household income. Given the well-documented problems with estimating income or expenditures from households surveys administered in developing countries, this issue must remain unresolved for the time being.

One could argue that the most important result of this analysis is the role played by two particular cultural values that appear to strongly influence community participation. The most striking of these is that respondents who exhibit little support for the notion of patron-client relationships (VALUE_D) are substantially more likely to participate in community activities. The log odds ratio (0.74) associated with rejection of patron-client relationships is significant at the 95 per cent level and, according to the odds ratio, those who reject patron-client relationships are more than twice (2.1) as likely to participate in community activities than those who support patron-client networks or obligations. In other words, this result says that rejection of patron-client relationships is possibly the most important contributor to community participation among these respondents.

In general, social relationships among people in Thailand have traditionally been based on reciprocal bonds or obligations that assume that the patron will undertake to support and assist his or her 'clients' in return for their loyalty and effort. As such, individuals who reject the importance or the underlying rationale of patron-client linkages might also be individuals who are less traditional or conservative in their belief systems, perhaps more educated and more likely to question existing modes of behaviour. It is reasonable to find that an untraditional individual in terms of beliefs would also be more likely to participate in community action, a non-traditional behaviour.

Furthermore, rejection of the value of individualism (VALUE_I) is associated with the increased likelihood of participation in community activities. The

---

[12] Duesenberry (1967) in his reformulation of the theory of saving (Chapter 3) argues that low economic status is felt more often and more keenly the more often someone makes a unfavourable comparison between the quality of goods she consumes and the goods consumed by others. The frequency of this comparison will depend on the ratio of her expenditures to those of others in her surroundings, which implies that those who are most dissatisfied with their income will be in the lowest income group.

coefficient of -0.91 has a more than 99 per cent level of significance although the odds ratio, Exp (B), indicates that rejecting the value of individualism makes it only slightly more likely (about 40 per cent) that an individual will participate. Traditionally, Thais have a deep-seated belief in the value of individualism which many argue is one reason citizen participation in governance is so limited even among the middle class. Someone who supports the cultural value of individualism believes that individuals do not have an obligation to perform work or support institutions beyond an individual's immediate interest.[13] Consequently, as hypothesized earlier, it makes intuitive sense that those who do not support this traditional Thai value would be more likely to engage in community-oriented behaviour and commit themselves to groups and institutions with communal objectives.

Other findings of the logistic models include the information that a number of socio-demographic variables, such as education, income (as estimated by expenditures), and length of tenure, are significant predictors of behaviour in many of the models. Education, for example, is a significant predictor of multiple environmental behaviours, including the presence of a metered water connection, boiling or filtering of drinking water, buying bottled water, disposing of waste properly in a bin, visiting the doctor frequently and taking children regularly to see a doctor. Similarly, income is a significant predictor of multiple behaviours, including the presence of a metered connection, buying bottled water, and deciding to go to a doctor when seriously ill.

Increased tenure, or the length of time a household has resided in a slum community, is associated with higher levels of community participation and an increased likelihood of visiting the doctor. Many researchers have found that security of tenure is a well-known determinant of household investment and such investments may include improving the community around one's residence as well as taking better care of household health through visiting the doctor (as well as, of course, becoming more familiar with local doctors and clinics; also Jimenez, 1987). At the same time, however, increased tenure makes it less likely that a household will buy bottled water. This may be because many of the older slums are located in the inner core of the city and have access to municipal piped water, which some believe to be uncontaminated and pure enough to drink straight from the tap. It may also be true that the longer one lives in a particular community, the less one worries about the water because, over time, one becomes accustomed to the conditions of one's environment. Most people in the slums of Bangkok are not aware that water quality is inconsistent across seasons and over time, and is deteriorating quite significantly. On balance, tenure is clearly an important explanatory variable in terms of predicting action and should be central in policy-maker strategies to improve environmental conditions.

---

[13] See ten Brummelhuis (1984) for a much more complete discussion on the nature of Thai individualism.

Another point worth noting is that the gender of the respondent seems to make relatively little difference to most of the analysed behaviours. Gender proved significant only in the case of proper disposal of solid waste in a bin or dumpster and the practice of taking children regularly to the doctor. In both cases, women were significantly more likely than men to engage in the behaviours. It may be that gender differences are not significant in terms of water treatment and community participation or it may be that respondents answered behavioural questions on behalf of their spouses or households as well as themselves. For example, when asked whether or not he boiled the household's water before drinking it, a male respondent, who is aware that the women in the household boil the water, may well have responded correctly for the household but not for himself. Unfortunately, the survey was not designed to explicitly explore the variations in male and female roles in the determination of behaviour, and that issue consequently merits further analysis.

Community Participation and Environmental Governance

To summarize, the survey of low-income urban residents in Bangkok indicated both positive and negative dimensions in terms of access to public services and environmental practices. Descriptive statistics indicated that, in terms of individual or household access to public services and environmental practices beneficial to the surrounding community, most respondents had access to private piped water, had legal and working water meters, and disposed of solid waste in municipal bins provided at the neighbourhood level. Almost half of the respondents were aware of community-based organizations in their neighbourhoods, and about one-third had participated in community activities over the previous year (most centring on neighbourhood clean-ups). Moreover, respondents believed that individuals, communities and municipal and BMA government officials could all work to improve environmental conditions in local neighbourhoods.

However, there were also behaviour, attitude and cultural value findings from the survey data that indicated that environmental improvement through community participation among slum dwellers might be problematic. Almost three-quarters of the respondents reported that their wastewater was dumped untreated into open sewers, almost one-third did not visit a doctor over the previous year when ill, and approximately one-third had experienced diarrhoea or gastroenteritis during the previous month but over half the sample saw no relationship between gastroenteritis and water quality. There are clearly problems in access to medical services and public understanding about the connections between environmental conditions and health that must be addressed by policy-makers. Respondents also indicated that contrary to the literature documenting the environmental problems in Thailand, which usually focus on traffic, noise and air pollution, insects and vermin constituted the largest problem for low-income slum residents (compared to the second-largest problem of bad odours, rubbish and noise stated by one-quarter of the sample).

Our multivariate logistic regression models further clarified these relationships among socio-demographic variables, cultural values, attitudes and behaviour. By exploring a variety of environmental practices, including individual treatment or purchase of water, sanitation behaviour and wastewater practices, the survey results indicated two potential, but perhaps conflicting, policy implications for improving environmental conditions and household health. The models indicated the central role of improved education and income levels among Bangkok's urban slum dwellers, especially with respect to addressing environmental deterioration and public policy in the Bangkok Metropolitan Region. Education and income levels were positively related to having a legal water meter, boiling drinking water, buying bottled water, placing waste in BMA neighbourhood bins, visiting doctors when ill and taking children to doctors when they were ill. Elevated socio-economic status clearly has a role to play in improving not only individual household health, but also, as shown by the models, proximate environmental quality.

But elevating socio-economic status for Bangkok's slum dwellers, although necessary for improving individual household health and well-being, may not encourage community participation in the prevention and resolution of local environmental degradation. Indeed, our models indicated in contrast that living longer in slum neighbourhoods and a lack of belief in patron-client relationships were related to a greater likelihood of community participation. The notion that living longer in a community may provide greater incentives to participate in improving local conditions is not surprising, since longer-term residence is central to notions of 'community' in both the industrialized and industrializing world.

The lack of belief in patron-client relationships, however, may point to a complex relationship among community participation, socio-economic status and cultural values. If patron-client relationships remain vital avenues by which Thai residents improve their socio-economic status, gain access to resources and influence, and eventually accumulate personal and family wealth, then the rejection of such a belief would suggest that such individuals are effectively excluded from the environmental and household health improvements associated with income and education discussed previously. However, the rejection of this ubiquitous social norm also indicates that such individuals may harbour notions about 'community' rarely practiced in the 'individualistic' Thai culture identified through our value measures of individualism, avoidance and immediate gratification. That is, in rejecting the patron-client relationship, such respondents indicate that they are less likely to believe in hierarchical social relationships, and therefore, may be more likely to believe in collective behaviour for environmental improvement. Thus, the very fact that such respondents are materially and cognitively excluded from the *status quo* of patron-client relationships may enable them to participate in community-based efforts that benefit households other than their own because they see a benefit to collective rather than individual action. For those households disenfranchised from the patron-client structure of Thai society, community-based mobilization and organization may prove a vital strategy, not only for improving environmental conditions and household and community health, but also, eventually, for gaining access to public and medical services.

Given this complex relationship, it might be worth attempting to frame environmental policy proposals in such a way that they encourage community participation at the same time as they subvert, to some extent, patron-client relationships. Patron-client relationships might, in that sense, become a dependent variable as well as an independent variable in the design of policy. Perhaps later efforts of such a line of research might try to treat these two variables, participation and support for patron-client relationships, as simultaneously determined if environmental and other policies actually succeed in breaking down patron-client relationships. Ultimately, policies that encourage community participation may be the key instrumental variables in the promotion of more participatory citizen-government relationships that many argue will need to supplant more traditional forms of governance. Additionally, encouraging community participation may also help transform the positive relationship between improved economic status and patron-client relations that we note above. This may be a line of thinking worth pursuing given that the general trend towards economic modernization in Southeast Asia is eroding some of these traditional routes to wealth formation.

The findings of this analysis also highlight the distinct role played by cultural values and attitudes in shaping and predicting important environmental behaviours such as community participation, health behaviours, treating water and disposal of solid waste. There has been limited scholarly work documenting the linkages among values, attitudes and behaviours in developing nations, and thus, the results presented in this chapter have been somewhat exploratory. However, given the findings presented here, the changing character of cultural values and attitudes in varying populations responsible for and affected by environmental practices needs to be studied further to understand their linkages, and their possible role in policy formation and implementation.

Additionally, following from this issue and on a methodological note, the type of descriptive analysis presented in this paper is not a substitute for careful economic and policy analysis of proposals to improve sanitation, water treatment and solid waste disposal. As an example, our work strengthens the foundation laid by Dale Whittington and others for the use of contingent valuation surveys of willingness to pay for investments in environmental programmes and infrastructure in low-income communities.[14] The question-naires used in such surveys should probably contain questions regarding environmental priorities and attitudes as well as individual or household willingness to pay for carefully described and specific hypothetical programmes. This kind of research would further extend the study of linkages whose importance we stress in this chapter.

At the same time, it is not possible to do serious economic evaluation without the kind of detailed information on current systems and resident perceptions presented here. Too often, environmental policies and sanitation systems are put in place because information is not available on household needs and perceptions, and yet households are expected to cognitively and politically accept them, and further, are expected to pay to support their maintenance and monitor their implementation. With the complexity indicated by our analysis, and the lack of success of infrastructure policies across the developing world, we believe that as a starting point for developing environmental management policies, the kind of information presented in this chapter on household attitudes and values surrounding environmental issues should be collected, analysed and integrated into any environmental strategy development. The significance of cultural values and attitudes indicated by the models presented imply that without adequate policy attention paid to such issues, environmental management policies, especially those which claim to incorporate community participation, are unlikely to succeed.

In addition, while economic and policy analysis can provide quantitative evidence for the development and implementation of infrastructure and other environmental policies, environmental governance in the twenty-first century will also be likely to focus on community participation to compensate for lessening Thai state intervention. Thus, the insertion of individual and household attitudes and perceptions, cultural norm structures and potential and actual behaviours by Bangkok's residents is a critical component of future environmental plans and policies. We should note that the survey data presented in this volume cannot take the place of careful qualitative research, which would provide analyses of the 'ways in which development interventions and market transactions become part of a longer, sedimented history of a place and its linkages with the wider world' (Bebbington, 2000, p. 495). However, these data do provide a first step in documenting the views and actions of those most excluded from environmental policy design, implementation and enforcement.

Indeed, because of the lack of resources available to these low-income slum dwellers, another vital dimension of environmental governance in Thailand

---

[14] See Choe et al. (1996) for an example from the Philippines or Whittington et al. (1991) for a case study in Nigeria.

will lie in the realm of non-governmental organizations. It is perhaps this arena that provides an even greater potential for environmental management and regulation becoming and remaining a permanent fixture outside of the central Thai state. It is to the role and impact of non-governmental organizations that we now turn.

Chapter 5

Environmental Governance and the Shadow State in Thailand

Within the countries of Southeast Asia, there exists a broad spectrum of non-governmental organization (NGO) activity. It ranges from very limited NGO activity in Burma and Singapore to limited but growing activity in Vietnam and Malaysia to very high levels of activity and organization in countries like the Philippines, Thailand and Indonesia (Clarke, 1998). In addition, a number of authors document significant NGO activity in newly industrializing East Asia, for example in Taiwan, Hong Kong and South Korea (Lee and So, 1998). Furthermore, it is widely agreed that in countries with low levels of NGO activity, the state is able to successfully control political mobilization and associational life, thereby minimizing the institutional space available for NGOs (Clarke, 1998, p. 39), and the potential for the development of non-state organizations resistant to the state and private capital interests.

The Philippines represents one of the best examples in Southeast Asia in terms of the influence that NGOs can have upon existing political institutions and policy formulation and implementation. The Philippines has the broadest network of NGOs in the Southeast Asian region. The success of Filipino NGOs at shaping the political agenda has also been linked to the way that the Philippine government managed the transition away from the Marcos dictatorship to a more participation-oriented government that naturally included and facilitated the active participation of grassroots-type NGOs (Magno, 1998). NGOs in other NGO-prevalent states (including Thailand, Indonesia, Taiwan and Hong Kong) have also been able to influence political discourse although they have, in general, had less of an impact on the actual practices of the state and corporate interests.

In terms of environmental governance, the discussion in Chapters 3 and 4 makes clear that the Thai state bureaucracy is central in the definition and response to environmental problems that increasingly face Thai civil society. But in addition to the Thai state, non-state actors, particularly non-governmental organizations (for example, local community-based grassroots organizations, resident organizations connected to regional and international efforts, colleges and universities, and international donor agencies) have played an uneven but increasingly visible role in contemporary Thai efforts to manage environmental problems and implement potential solutions. While the form and function of the Thai state, and especially how these relate to environmental management, have been explored by a number of scholars, there has been much less investigation of non-state activities in environmental resource management in Thailand. To understand the complex nature of environmental governance, the role of non-state actors and organizations must be more fully explained.[1]

In understanding the role of non-state actors in environmental management in Thailand, we assume that non-governmental, voluntary and community-based organizations do not necessarily comprise an organized set of actors in terms of environmental policies and local practices. On the contrary, the non-governmental voluntary sector instead represents a fragmented universe of interests, based on geography, political affiliation, ethnicity and other social identities and ties. On the one hand, even with this fragmentation, non-governmental organizations and groups involved in environmental management and policies may constitute a 'shadow state', supporting and reinforcing state or corporate interests while acting on behalf of local communities and disenfranchised populations. On the other hand, this fragmented universe of interests may be critical to developing less paternalistic, and less patron-driven structures, especially given the central integration of the Thai elite and military classes in Thai state structures, consequently heightening the potential for resistant practices of governance in Thailand. That is, effective and influential non-governmental actors and organizations may also offer the potential for a resistant governance structure, acting to incrementally and radically address class-based oppression through environmental action.[2] Class-based resistance remains highly problematic in environmental governance because such efforts tend to be focused on local communities, may not address regional and national inequities, and may be subsumed in larger regulatory practices designed to align local participation with individual entrepreneurialism (Martin and Ritchie, 1999). Even so, environmental action at the local level may represent opportunities for overt and covert acts of resistance. Environmental action could represent direct critique of the standard central government and industry position toward urban environmental issues (and the populations exposed to pollution and contamination) or indirect opposition through community-based (and therefore non-state

[1] David Gibbs and Andrew Jonas (2000, p. 300) argue that environmental policy design and implementation, especially at the local level, has shifted somewhat from state-centred to non-governmental arenas: 'Increased emphasis is placed on developing local partnerships, networks, and other extra-governmental institutions that are considered appropriate for delivering sustainable and local approaches to the environment.'

[2] Local uncoordinated acts of resistance by individuals and groups, even those that are covert and everyday, may have broader impacts. Scott (1985) argues for example that small covert everyday acts, such as seeming ignorance and sabotage, may be or become part of a larger set of culturally defined actions that serve to conceal resistance to dominant and unjust systems and social relations.

centred) activities aimed at improving the quality of life for the urban poor (Mullings, 1999; Pile and Keith, 1997).

To investigate these issues, this chapter explores the parameters of a possible Thai shadow state and the potential for resistance by non-governmental organizations through environmental governance by deciphering the form and potential effectiveness of varying groups and organizations in Thailand. The chapter first provides a conceptual overview of our underlying premises concerning the potential existence of a shadow state, supporting Thai state interests, and the countervailing possibility for an environmental governance of resistance by NGOs. It then describes the historical development of Thai NGOs, tracing the importance of particular individuals and organizations in the embryonic and contemporary development of Thai non-governmental organizations and institutions. Next, the chapter turns to the specific NGOs that focus on environmental issues, to analyse the potential for a shadow state, and its potential for supporting or resisting state and private sector interests. Finally, the chapter concludes by outlining broader conceptual issues, building on this historical context and scholarly thinking about environmental governance.

Non-State Actors: Shadow State or Resistance?

International donor agencies, in their efforts to improve the efficiency and effectiveness of infrastructure policies in developing countries, have increasingly emphasized in their funding and analyses a transition in governing from government policies and methods to 'an extension of individual obligations' (Raco and Imrie, 2000). This emphasis on individual contribution to communal quality of life is a reaction against inefficient and ineffective central governmental policies and structures, instead relying on communitarian responses to local environmental challenges. While this adjustment in thinking by international donors offers an alternative to what has in practice constituted the centralization of power and resources among the governmental, industrial, and military elite in Southeast Asia, challenges to power realignment remain. That is, a renewed focus on decentralization, community participation and individual obligation, while seemingly preferable to centralized and seemingly ineffective government structures, may do little to overcome structural and systemic characteristics; such political economic structures serve to reinforce social relations that create uneven environmental problems across Thai neighbourhoods, with much of the negative environmental burden concentrated in low-income, tenure-insecure urban communities (see Rose and Miller, 1992). In addition, from a pragmatic point of view, as environmental governance becomes more indirect (that is, more dependent on NGOs for policy design, implementation and enforcement), it may be more problematic to implement environmental management policies and ensure that the objectives of these policies are met (see Salamon, 1995, p. 27).

In considering the possible state-supportive or resistant character of Thai non-state actors in environmental governance, we build on conceptual and empirical studies of the third sector based largely in the developed world. We envisage the 'shadow state' in Thailand's environmental governance as paralleling to some degree the form and function of the 'shadow state' in industrialized countries. Here, we draw on the work of Jennifer Wolch (1989, 1990) and others who have argued that in the US and other Western nation states, during the 1980s, welfare state restructuring resulted in a greater reliance on privatization and the non-profit sector for population welfare and social service provision. Wolch, in particular, argues that the shifting respon-sibility for population welfare from the state has resulted in the formation of a 'shadow state apparatus.' She describes a shadow state as 'a para-state apparatus with collective service responsibilities previously shouldered by the public sector, administered outside traditional democratic politics, but yet controlled in both formal and informal ways by the state' (Wolch, 1989, p. 201). This para-state apparatus may have as its 'street-level bureaucrats' private sector individuals such as 'bankers and businessmen, hospital administrators, and corporate tax accountants' who implement the policies and programmes promoted by state bureaucracies (Salamon, 1995, p. 22). In this sense, such street-level bureaucrats become unmonitored extensions of the state, implementing state-sponsored programmes while minimizing potential critique by the populace.

This shadow state apparatus consists of 'voluntary organizations with collective responsibilities, which ... are strongly affected by state resources and constraints' (Wolch, 1989, p. 198). Wolch's work suggests that instead of constituting interests clearly in opposition to state policies and ideologies, non-profit and non-governmental voluntary and charitable organizations in the US may be acting consciously or unconsciously to promote social stability through social, cultural and other types of service provision, and consequently, serve to support the ongoing mission of state bureaucratic agencies (that is, social stability). The control and influence by the state over the non-governmental voluntary and charitable sector (thereby creating an additional 'apparatus' of the state) stems, in the US, from tax-exempt status granted to such organiza-tions and the funding mechanisms emanating from governmental agencies and non-profit foundations.

Given such regulation and funding instruments, the activities of non-governmental and charitable organizations are determined in large degree by state-mandated policies, or by non-profit foundation priorities, where such funding is tied to specific programmatic outcomes targeted at particular population groups (Takahashi and Smutny, 1998). The shadow state may therefore function to enable social control mechanisms through the privatization of state-initiated policies and plans, thereby facilitating *status quo* social conditions while minimizing public sector participation and oversight. The minimization of clear public sector involvement may, according to Jennifer Wolch, lead to further penetration into social life by the state, as non-state organizations function to further state-sanctioned, -funded and -initiated actions. Further, such non-state action may subsequently lead to greater 'management' of the populace through activities such as government-sponsored programmes implemented by non-governmental organizations (Wolch and deVerteuil, 2001). That is, through the provision of services, non-governmental agencies may effectively further state goals for social stability and state dependence, dampening the potential for resistance and mobilization.[3]

To enable such non-state actors to further state objectives, the state actively or indirectly devises and uses strategies that maximize appropriate behaviour. As Martin Jones (1998) has argued, one of the important questions in theorizing governance (and we argue for environmental governance as well) is the extent to which the state works in 'mobilising support for its policies and ideologies, suppressing those interest groups which pose a threat, and lastly, *institutionalizing* support' (p. 986). However, the state may also see the non-governmental sector as monolithically threatening to the stability and sustainability of state power. Thus, instead of the state using various mechanisms to encourage and mandate non-state organizations to conform to state goals, the Thai state through the form and function of its environmental apparatuses has at times acted to effectively silence and repress non-governmental groups and organizations. Environmental governance policies and actions in such situations are directly expressed by state-centred policy design, implementation and enforcement with limited interaction and interference by non-state groups and organizations.

There are of course many degrees and types of interaction between the state and NGOs that fall between extreme repression and complete cooptation.

---

[3] Similar arguments have been made about the utility of seeming empowerment and loosening of monitoring in public sector management. Martin and Ritchie (1999, p. 120, citing Hoggett, 1991) argue 'High trust relationships between workers in participatory groups are instrumentalised by devolving particular responsibilities, and drawing on the communicative solidarities of the group to regulate collective action towards management goals. Ironically, "empowering" social competencies such as communication, collective planning and group skills are developed and harnessed towards "productive" ends whilst strategic power is centralised with management ... The instrumentalisation of social relations of labour can occur when the interests of the worker are perceived to coincide with management. These situations involve a reconfiguration of regulation to local sites rather than "deregulation".'

Costen (1998), for example, defines six possible types of relationships: repression by the state of NGOs; competition between the state and NGOs (for example, for jurisdictional influence); contracting (for example, the privatization of public services); cooperation through coordinated programme delivery, information sharing, or leveraging of resources; complementarity (for example, where state and NGO efforts provide distinct yet coordinated services); and collabouration (for example, public-private partnerships).

Korten (1990) outlines a generational theory of NGOs that a number of authors use to describe Thai NGO development. In essence, as NGOs age and become more established, their strategies and the arenas in which they exert pressure shift from the bottom to the top. Pratt (1993, p. 12) notes that 'The development of strategies of the majority of NGOs working with the urban poor in Thailand correspond primarily with Korten's 'first generation'.' Pratt later adds: '[w]ith a shift toward 'third generation' NGO strategies, however, there may be a need for more systematic channels of accountability' (1993, p. 17). The next section provides an overview of the development of NGOs in Thailand, reflecting varying degrees of cooperation and conflict between Thai NGOs and the central state, and indicating a trend toward greater corporatism throughout the non-governmental sector.

From Resistance to Shadow State: The Development of Thai Non-Governmental Organizations

The roots of NGOs in Thailand, as in many other places, originate in religious institutions. For example, for centuries, Buddhist monasteries supplied assistance to ill and impoverished individuals and families as well as basic public education (at least for boys and young men), and provided convenient and safe sites for communal functions and meetings throughout the kingdom. These NGOs had their philosophical roots in the work of a Buddhist scholar named Buddhadasa Bhikku, who emphasized the duty of Buddhists to alleviate suffering, as well as the ideas of Pridi Bhanomyong, a bureaucrat who wrote in the 1930s, who believed in the reforming potential of bureaucracy (Pasuk and Baker, 1995, p. 348). In addition, since the early Bangkok period, Christian missionaries were active in Thailand, building and maintaining hospitals and schools as well as providing basic needs to some of the most marginalized groups in Thai society, including the hill tribes and slum dwellers.

The continuing, and not insignificant, efforts of religious organizations in Thailand were supplanted by an impressive number and variety of NGOs. As in a number of other Southeast Asian States (or, more generally, the rest of the world), NGOs have emerged as a very important force in shaping political discourse and influencing policy at a variety of levels across different spheres of activity. The first modern forms of philanthropy in Thailand included the founding of Sapha U-nalom Deaeng (or what later became the Red Cross) and the Boy Scouts, which came into existence during the reign of King Rama VI (1910–25). Other philanthropic organizations emerged at about this same time, many of which were involved in the provision of social services and some of which embraced quite liberal political agendas. A typical example of NGOs during the early twentieth century included the Chinese secret societies that were transformed into mutual aid and speech associations to address the needs of occupational groups and new immigrants.

It was at this point that Thai NGOs began to represent a potential political force or sector. In response to the sector's growing size and presence, which coincided with the Depression, the Thai government exerted significant control over what it perceived to be a growing and potentially threatening group of organizations similar in purpose and ideological position. Faced with the proliferation of added diversity within the non-profit sector, the central government ordered associations to register under the National Cultural Act of 1942. The implementation of this Act resulted in the direct monitoring and control over the activities of such organizations since the Act mandated that all activities had to be non-political, allowing only social and philanthropic endeavours. Thus the response of the Thai state to NGOs during this period of NGO expansion and diversification can be characterized as antagonistic, suspicious and controlling until the early 1970s.

More recent changes in the non-governmental sector, particularly environmental Thai NGOs, can be divided into two temporally distinct periods: the 1970s/80s and the 1990s and forward. In 1973, Thailand was shaken by a *coup d'etat* led by university students who espoused a decidedly left wing and participatory or democratic philosophy. Development NGOs proliferated rapidly between 1973 and 1976, during the period of openness when the government was run by freely elected non-military members of parliament. Some of these students experienced their political and social formation as workers at summer camps such as those run by the student-initiated Thailand Rural Reconstruction Movement (TRRM), the country's first NGO emphasizing development, that ran camps for upper-class youths to help familiarize them with rural poverty and economic disparity. The TRRM was founded by a bureaucrat, Central Bank Chairman Dr. Puey Ungphakorn, in 1969. The organization of these students, along with many other middle- and upper-class Thais, constituted the catalyst for the formation of a number of new NGOs that did not formally register with the government. These organizations demanded what were seen as radical reforms to stop the transfer of national resources from the poorer to the wealthier groups of Thai society (Ponsapich, 1996).

To some extent, political economy can help explain the rapid proliferation of non-governmental organizations in both Thailand and the Southeast Asian region as a whole. In Thailand, as in Indonesia and other countries, class-based social movements that started in the 1970s gradually disintegrated into a variety of issued-based groups that focused on a myriad of concerns including democracy and human rights, government economic strategy and agrarian and rural development issues such as land reform, minority rights and the environment. NGOs in Thailand were subsequently repressed and became inactive following the relatively forceful military coup of 1976, when the democratically elected coalition of left wing groups led by the Communist Party of Thailand (CPT) disintegrated following well-organized counter-insurgency efforts. The new military government took almost immediate action against many of the Thai NGOs. The primary method of attack was to label different organizations as community threats and to force these groups underground or, more commonly, into the Thai jungle. It was not until the early 1980s, during the more liberal government of General Prem Tinsulanond, that NGOs were able to restart their development efforts, when those who had fled into the jungle were invited back through the promulgation of a general amnesty. In present-day Thailand, NGOs are often guided by former CPT activists who maintained their commitment to social change and yet had to find another means to work for improvements within the Thai political system.

Another explanation, advanced by Clarke (1998), for the growth in NGOs in Southeast Asia lies in the gradual replacement of social relationships characterized by dyadic ties with more formal and varied forms of social organization. NGOs and other public groups have begun to replace more traditional social organizations such as the *gotong royong*, the traditional spirit of mutual help in Indonesia, or the traditional funeral societies in Thailand. In particular, some Thai NGOs permit prominent intellectuals to maintain traditional Buddhist values despite the contradictions posed by rapid economic development and concomitant social change. In addition, the significance of NGOs within the Thai political system stems in part from the nature of the Thai state and what Clarke (1998) describes as the 'institutional vacuum' created by Thai political relationships. According to Unger (1998), and others, and consistent with the description of the Thai state developed in an Chapter 3, the Thai political system is characterized by weak ideological cohesion, frequent party-changing and low membership. These features reduce the capacity of political parties to contribute effectively to resolving important political issues and problems. In essence, most parties within the Thai political system are incapable of representing the increasingly diverse interests that comprise Thai society and are unable to include new social groups into the political arena. Small, flexible and relatively well-organized NGOs, in contrast, provide some of these marginalized groups with access, though limited, to political debate and opportunities for policy change. As such they are well positioned to help fill the vacuum created by the Thai state.

As amnesty lifted the control and repression of NGOs, a number of the dormant and unregistered groups began to revive their activities and other new groups formed. A variety of development-oriented grassroots organizations emerged and became significantly more active over time. By the early 1980s, non-profit NGOs were assuming a small but important role in the development of Thai society despite some remaining limitations. Most public-spirited development-oriented NGOs established during this era were not registered, consisted primarily of middle-class members and operated from the belief that NGOs provided an alternative means to tackle development problems. These NGOs, however, were composed of many workers who in the previous period were suspected of being competitors and antagonists of state policies and officials. Indeed, being somewhat anti-bureaucratic in formative experience, NGO workers also had negative attitudes toward government officials. Thus, in the early 1980s, there was a significant level of tension between NGOs and governmental bureaucracies. Only more recently, since the early 1990s, has there been dialogue, exchanges and cooperation between the two sectors. This is evident in the establishment of the Joint Coordination Committee in 1992, between governmental bureaucracies and NGOs where representatives of government agencies and volunteer groups meet regularly.

Thus, the NGOs that formed in 1970s and 1980s developed during a time of intermittent hostility on the part of the government towards NGOs. These NGOs, which often had political as well as substantive-based agendas such as the environment, poverty alleviation, health improvement or agricultural assistance, were characterized by a strong interest in participatory learning and development and grassroots mobilization. Since the 1980s, Thai NGOs have been particularly active in the political arena. For example, following the abolishment of parliament by the military in February of 1991, NGOs became a small but important source of opposition to the relatively repressive regime of the National Peace-Keeping Council (NPKC) headed by General Suchinda. NGOs undermined government plans to introduce a new and less democratic constitution in 1991 and organized frequent and large-scale demonstrations against the NPKC government. A particularly effective NGO was PollWatch, which was composed of thirty-three NGO leaders who came together in January 1992 to help monitor elections scheduled for March 1992. According to Callahan (1998, p. 104), 'In the three months leading up to the March 1992 elections, PollWatch recruited 20,000 volunteers from all over Thailand. They had two tasks: to curb vote-buying and to encourage democratic conscious-ness.' In this effort, PollWatch was at least partially successful in curbing voting irregularities. Many authors stress the major role that these NGOs played, and continue to play, in the consolidation of democracy in Thailand:

> Since the heady days of 1992 ... there has been a gradual process of democrati-
> zation, as the emphasis of NGO political activity shifts from the 'high politics',
> systematic opposition to authoritarianism centered on broad coalitions and sus-
> tained mobilization, to 'low politics', an emphasis on discrete issues by smaller
> issue-based coalitions. Yet, Thai NGOs continue to fulfill roles which contribute
> to the consolidation of democracy (Clarke, 1998, p. 137).[4]

This type of modern NGO was, in the 1990s, joined by a second group of issue-based NGOs that are sometimes referred to as 'business NGOs' (Lee and So, 1999). These NGOs are so characterized because they are often operated and funded by corporations; their methods (outreach through advertising and public education) are much less confrontational and more collabourative in nature in terms of interactions with the Thai state. This newest type of NGO is particularly visible in the environmental arena, causing Lee and So (1998, p. 294) to go so far as to label Thailand as moving along the 'corporatist path' based upon their belief that the major proponents of its environmental move-ment currently are corporations or corporate-affiliated or dependent NGOs.

The pragmatic need for public services and the lack of public funds may make such corporatist NGO organizations in addition to more state-resistant non-governmental groups critical to the ongoing economic and social vitality of Thailand. With the economic downturn in Thailand and across Southeast Asia, the non-governmental provision of services may expand in significant ways. There are various reasons why this might occur. Non-governmental service provision may be seen as more efficient than state bureaucratic efforts, especially with respect to corporate-oriented NGOs. Because of this cost-efficiency, public services may be expanded to cope with expanding and diversifying needs (NGOs may be more flexible, innovative and adaptable than state organizations in providing public services; for example, Eade, 2000; Higgott et al., 2000). Consequently, the Thai state (and international donors) may want to depend more on non-state actors, particularly those that serve Thai state interests, to absorb the costs of environmental management to maintain economic growth and stability.

In particular, one might expect that NGOs that focus on environmental issues might be relatively affected by recent economic and political shifts in Thailand. The lack of available resources available to local and regional governing bodies, the calls for more participation by urban residents from international donors and rapidly deteriorating conditions within major cities point to a context where Thai NGOs with aspirations around improving urban

---

[4] Clarke refers here to the popularly-led demonstration in 1992, which unseated an military-backed coalition from power. The 1992 demonstration, or people's coup, marked a new era, to some extent, in Thai politics in that the mainstream urban population of Thailand took concrete action to increase the democracy and responsiveness of the Thai state.

environments might find support from diverse sources. Both urban residents and international donors, for example, are likely to be more willing than in the past to seek to work with environmental NGOs that focus on providing both cleaner communities and on increased public voice. To explore the supportive and resistant role of Thai NGOs, we turn now to a more detailed discussion of NGOs in Thailand especially concerned with environmental issues.

Environmental NGOs in Thailand: Toward a Corporatist Shadow State

In Thailand, estimates suggest that as of 1995 there were more than 10,000 non-governmental organizations, an increase of about 250 per cent from 1984 (Rodan, 1997). Thai NGOs, as elsewhere in Southeast Asia, have become more significant players not only because they have grown in size but also because they have become an important conduit for overseas development assistance and have used, or are using, this function to carve a larger political and service-oriented role for themselves. Reliance on NGOs in Thailand has only recently become significant. As discussed previously, since the late 1980s, social organizations within Southeast Asia as a whole have expanded exponentially in number and influence. Prior to the economic downturn, the stable economic growth experienced by the region as a whole and efforts to engage in political reform, in particular, benefited business associations and stimulated their involvement in public policy (Laothamatas, 1992). Such corporatist NGOs tended to support the Thai state's interests in economic growth. In the last decade, NGOs have also appeared as critical new institutional actors and their rapid growth has echoed the 'associational revolution' evident in many places within the developing world, creating potentials for a more resistant component of the Thai non-governmental sector. As noted by Clarke (1998) this poses 'a major challenge to traditional conceptions of the voluntary sector in the regional political science literature.'

Environmental NGOs in Thailand have had a mixed record of success, especially in urban areas. While at some level, they have helped communities to improve their living situations, NGOs and grassroots groups have, at another level, been thwarted in their efforts to empower communities. Uneven development and systemic lack of access to resources have left such NGOs and their local communities without access to significant power to alter substantially public service provision. Many low-income communities in Bangkok remain voiceless and subject to almost intolerable conditions of contamination and filth. Despite living in a rapidly developing and modernizing capital that until recently experienced tremendous (10 per cent or more) rates of economic growth, poor urban residents are not participating in nor benefiting from improved environmental policy.

In general, environmental NGOs in Thailand can be classified as issue-based groups or non-profit groups that are occasionally and loosely organized under an umbrella organization. Umbrella organizations can be registered or non-registered and, according to Ponsapich (1996), there are fewer than twenty registered umbrella organizations in Thailand. Most of these registered umbrella groups are involved in religious or vocational activities, although about one-third of the umbrella groups are interested in issues related to social welfare. In the main, however, most NGOs that focus on specific issues and are development-oriented, including the majority of environmental NGOs, do not choose to register. In this aspect Thai environmental NGOs differ from some of the other NGOs in Thailand including, for example, the NGOs that are recognized by and legally registered with the Thai government, associations, labour unions and federations, and foundations. Most of these registered NGOs focus on funding cultural and educational activities although some also work on labour-related issues.

Thailand

Source: http://www.dpf.or.th/eng/projects/
index.html

Environmental NGOs, in contrast, comprise a large number of unregistered groups organized for specific purposes but without official legal status; these organizations and their activities are often referred to as projects, working groups or forums. In general, NGOs of this type are relatively small and dedicated to issues such as public welfare, community development, human rights, environmental protection and cultural promotion, that is, comprising a potentially weak group of NGOs resistant to state and private sector interests. These kinds of advocacy groups often do not register officially with the central government because of the arduous endowment and membership requirements imposed by Thai law. Thai law also requires registered groups to report regularly on their funding and activities, providing a further disincentive to legal registration. These types of unregistered NGOs occasionally combine efforts under umbrella organizations or coordinating committees.

The Seventh Economic and Social Development Plan (1992–96) helped to spark a loosening (to some degree) of control of the non-governmental sector.

In particular, some measures were taken to revise registration procedures and tax regulations to further promote NGO activities. In the main, however, most NGOs continue to remain outside the world of official recognition, providing them with greater freedom and less state monitoring, but also limiting them in terms of access to financial resources and political power. At the same time, the Thai bureaucracy passed modified tax regulations to promote private sector participation in development activities.

The Thai government's attitude towards NGOs is, at this time, relatively supportive even if the organizations are not registered. Thai leadership has recognized that both the NGO and private sectors can contribute significantly to economic and social-political development in Thailand and that both registered and non-registered NGOs may supplement and even partially replace the public sector in development-oriented activities. The Thai government apparently hopes to save financial resources by persuading the private sector to become more involved in development activities at the local level, which has occurred since the early 1990s. Because of the massive environmental and housing needs experienced by low-income communities throughout Bangkok, much of the focus of potentially resistant NGOs would remain on the provision of basic needs, shelter and public services. That is, NGOs would probably be burdened with coping with multiple needs, such as housing, employment and urban environmental degradation; thus, the available resources and time to devote to social mobilization and greater political participation would be highly constrained.

As in many Southeast Asian states, the 1980s and 1990s saw a rapidly increasing interest in environmental issues; Thailand has seen a concomitant rise in the number and heterogeneity of environmentally-based NGOs. Some NGOs concentrate simply on natural resource exploitation in rural communities while others are interested in promoting conservation of endangered land or species. Others, primarily in Bangkok or other large cities, focus primarily on the social dimension of the impacts of development particularly in terms of destruction of the urban environment. The range of NGOs in this latter arena include international NGOs such as the World Wildlife Fund or CARE, Thailand's own Project on Ecological Recovery (a vocal group that advocates environmental issues very actively), and small community groups intent on cleaning up their slums and or neighbourhoods.

There is no single organization that functions as a coordinating body among NGOs working on environmental issues. In fact, and similar to other spheres of NGO activity in Thailand, there is a significant division between Thai environmental NGOs defined by when they came into existence and the source of their political influence or motivation. The more senior group of NGOs 'continue to set up village-level organizations, organize regional or issue-based protests, represent local demands and support mass membership in people's organizations' (Clarke, 1998, pp. 47–8). Lee (1998), for example, notes that almost all the community improvement organizations in Thailand's urban slums were initiated or set up by either government officials or NGO workers.[5] Many of these NGOs were created in the 1980s and have expanded their strategies to include campaigning for public policy and legislative reform on issues such as human rights, freedom of information, environmental conservation and land reform, monitoring parliamentary procedures and placing stories with media groups.[6] One could argue that the social relationships created among NGO members and community participants are, in essence, collabourative and based on the notion of participatory action. The basic philosophy underlying the actions of such NGOs is to resist hierarchical decision-making, enabling as far as possible equal relationships among leaders, organizers and participants (Pratt, 1993).

An example of this type of NGO is the aforementioned Project for Ecological Recovery (PER), which was founded in 1985 by a community development activist named Witoon Permpongascharoen working in the province of Kanchanaburi. The PER developed from an environmental coalition of Buddhist monks, grassroots NGOs, students, rural residents and several environmentally-minded government officials who worked together to protest against the proposed construction of a very controversial dam known as the Nam Choan Dam. The stated mission of PER is to mobilize people to address specific environmental problems that affect their livelihood. Not surprisingly, the principal activities of the organization involve the training of and advocacy for local groups. According to Hirsch (1994) and Quigley (1996), the group advocates an anti-bureaucratic, egalitarian ideology born of opposition to the conventional development path and will not hesitate to adopt a confrontational strategy if needed. PER is credited by many for giving voice to local rural people who were not often included in the development of government policy particularly in the case of major development projects with enormous environmental impacts. The PER has a small permanent staff of about ten and an annual budget of approximately US$100,000, which comes primarily from international agencies and overseas organizations in the form of grants.

More recently, however, as mentioned earlier in the chapter, there has been a proliferation of business NGOs that has created a second type of social

[5] Lee (1998) further points out that there are major problems with such externally imposed organizational structures including: (1) such structures based on an elected community committee do not necessarily lend themselves to becoming a strong organization; (2) outside actors tend to depend too much on formal structures and inadvertently ignore groups outside of the formal system; and (3) these types of community organizations tend to be dominated by outside development agents.

[6] Pratt (1993, p. 12) makes a distinction between the activities of rural and urban community development NGOs noting that: 'Almost all urban NGOs in Thailand work on various types of slum community development activities: physical improvement (especially infrastructure), social services, economic improvement, and local empowerment.' The vast majority of Thai environmental NGOs are rural, however, and 'engage in a variety of sector specific activities'.

relationship between NGOs and their target communities. These relationships are based in NGO-industry collabourations and take advantage of the well-known and deep-rooted linkages between the political leadership and the business sector in Thailand (Bell, 1997). Such collabourations, according to Yamamoto and Ashizawa (1999), not only satisfy corporations' requirements for good public relations and fulfil their (somewhat limited) senses of corporate responsibility but also allow NGOs to avoid any protracted funding crises or heavy reliance on their own governments for necessary operating funds. Lee and So (1998, p. 281) suggest that there are three major categories of specific environmental activities supported by the private sector: sponsorship of environmental awareness campaigns, participation and support for environmental projects designed by environmental NGOs, and active response to the rise of green consumerism.

Examples of such activities, as indicated by Lee and So (1998), include the Thailand Environment Institute, based inside Bangchak Petroleum Co. Ltd, and Thai Environment and Community Development Association or 'Magic Eyes'. Bangchak Petroleum (directed by Sophon Suphaphong, one of Thailand's most well-known business environmentalists) helped to create the Thailand Environmental and Development Network (TEDNET) whose goals included the coordination of environmental action between government and NGOs. In addition, Bangchak is one of the primary financial backers of the Thailand Environmental Institute (TEI), a very visible policy think-tank, which came into being in the early 1990s only to run into financial and other difficulties following the untimely death of its charismatic and powerful director. Bangchak Petroleum presumably invests in these environmental efforts to respond to the negative image of industry that has been publicized by smaller environmentalist groups over the years and, of course, because of some sense of corporate responsibility regarding environmental protection.

Magic Eyes was created and sponsored in 1984 by Khunying Chodchoy Sophonpanich, the daughter of the founder of Bangkok Bank. The programme's symbol is a pair of watchful eyes, plastered on public dustbins cans throughout the city of Bangkok. The mission of the organization is to enhance cooperation among business, government and citizens in improving the quality of the environment through creating popular awareness by encouraging people to engage in small-scale efforts that improve local environments. Specifically, there are four general programme goals: (1) local cleanups, (2) waste reduction, (3) environmental appreciation, and (4) special initiatives. Some of the organization's concrete efforts include public education campaigns to reduce littering as well as a very successful effort to curb the use of Styrofoam in traditional festivals. In recent years, the organization has begun to engage in policy development with government authorities.

Magic Eyes is very well placed to influence policy, especially when compared to many small environmental NGOs. The organization is extremely well-funded and is the largest environmental NGO in Thailand with an annual operating budget of US$360,000 in 1997 and a full-time staff of twenty (So and Lee, 1998, p. 135). This funding is drawn primarily from its corporate members such as Thai Airlines, Sima Cement and Shell of Thailand. The organization is influential and well known because it has links to and knowledge of the media. A number of Magic Eyes' board members are marketing executives from Thailand's leading corporations. In addition, all the major television stations are represented on its board of directors and therefore Magic Eyes' activities are extremely well reported in the mass media. Its smaller advisory board has a significant number of current and former government officials, and Magic Eyes' officials also serve on a number of prominent government committees at both the national and local level.

Non-State Environmental Governance in Thailand

This discussion is not meant to imply that Thai NGOs control or even dominate political discourse and create or implement public policy. Rather, Thai NGOs represent a possible avenue for some individuals and interests, usually middle-class and educated, into the Thai state decision-making process. Since the early 1990s, NGOs have shifted in focus from systematic opposition to authoritarianism to an emphasis on discrete issues through smaller issue-based coalitions. It remains to be seen the degree to which such NGOs can actually affect broader participation in decision-making and can facilitate the restructuring of a very centralized and unresponsive Thai state (Eccleston and Potter, 1996).

More generally, Thai ideological and philosophical positions constitute substantial challenges to the effectiveness of NGOs in countering environmental degradation in Thailand. As indicated in Chapter 4, Thai society largely functions around a cultural norm structure emphasizing individualism, patron-client relationships and fatalism, contributing to a lack of engagement by local residents (that is, those individuals not inculcated in the Thai political elite and military strata) in community-based efforts seeking to improve and manage local environmental conditions (Daniere and Takahashi, 1999). This ideological focus not only on individuals and their immediate families, but also on the futility of community or social action, has been reinforced through Thai state policies and bureaucratic action. That is, Thai state and bureaucratic actions have done little to facilitate effective influence by most NGOs in Thai environmental policies, particularly grassroots efforts, and indeed, in specific eras has deliberately worked to eradicate their participation in policy-making. The convoluted nature of the Thai state as described in Chapter 3 provides an effective barrier to accessing relevant ministries and departments, but also remains a vital apparatus for intimidating low-income and disenfranchised individuals concerned with local environmental conditions. The results of such disenfranchisement manifest themselves in ways both conducive to and in opposition to varying state efforts. In other words, while the convoluted nature of the Thai state remains an effective obstacle to greater civic participation in decision-making, it also means that state decisions may be difficult to implement because of the fragmented and overlapping nature of the Thai state bureaucracy.

However, even in this difficult policy context, non-governmental voluntary and community-based efforts are seen as preferable to centralized government action by international donor agencies in tackling complex issues such as environmental management in rapidly developing, and now economically vulnerable, nation states such as Thailand. Ideologically and philosophically, international donor agencies have focused on community-based action as a potential antidote to hierarchical policy-making that actively excludes low-income, disenfranchised individuals and families experiencing political and social isolation. NGOs are seen as more pluralist, less top-down and hierarchical, and more inclusive, providing access by marginalized and disenfranchised individuals to arenas of political and institutional power (Bebbington, 1997). Questions, however, remain about the degree of inclusiveness, access, and pluralism actually existing in NGOs, and about their potential impact in counteracting the social regulation inherent in state support of NGOs (Farrington et al., 1993; Levy, 1996).

Not surprisingly, during the two different eras of environmental NGOs in Thailand, these organizations have had very different relationships with the Thai state. This is in part due to historical circumstances; during the 1970s and 1980s, NGO-state relations were often antagonistic primarily because many NGOs had their roots in groups first organized to resist the political *status quo*. However, in more recent years, institutional space has opened as environmental NGOs are increasingly perceived as organizations without a radical political agenda, and having ties to powerful corporations and the governmental and military elite. Therefore, the NGOs that have been created more recently have been exposed to much lower levels of state aggression, but at the same time these organizations may be more likely to be corporatist in nature, or at least to have collabourative private sector relationships.

Despite the different institutional learning environments of the 1970s/80s and the 1990s, and perhaps because of the different levels of willingness on the part of NGOs to work with the Thai state, all environmental NGOs are to some extent constricted by the rules of registration and operation that the Thai government imposes upon NGOs. As discussed by Pongsapich (1996) because of burdensome endowment and membership requirements, many grassroots and advocacy groups prefer not to register. Not registering brings the added benefit of not having to comply with state reporting requirements. Funding for NGOs is similarly constrained by the Thai government as, for example, all bilateral funding to NGOs must be coordinated by the Department of Technical and Economic Cooperation (Ponsapich, 1996) unless the funding entity or country designs and implements a special funding mechanism.

The controlling nature of the Thai state's relationship with NGOs has undoubtedly hindered the ability of NGOs to access international funding and to operate completely within the law; however, it has not prevented all NGOs from collabourating with the Thai state on specific projects, even NGOs that are outside the corporate and military elite power structure. There is some evidence of Thai state-NGO collabouration, such as the Urban Community Development Office, which attempts to coordinate NGO activities within greater Bangkok and includes so-called 'informal' NGOs on its board and staff. In addition, the Thai state in public documents (under the Seventh and Eighth National Social and Economic Development Plans) has articulated some willingness to collabourate with environmental NGOs. In the Seventh Plan (1992–96), the Thai government established the Environment Fund to help support projects concerned with environmental protection and natural resource conservation and to which NGOs are eligible to apply for funding. This explicit and targeted funding builds on recommendations of the Sixth National Plan, which envisioned 'targeting less advantaged areas and populations groups with reference to grassroots development':

> The Sixth National Economic and Social Development Plan ... aims to encourage the public to find solutions to their own problems on the basis of self-reliance: local organizations at the village level will be strengthened; local resources will be mobilized; and people will be encouraged to solve their own and community problems (Lee and So, 1998, p. 137).

Taken together, it appears that the Thai government is moving towards an expanded role for NGOs, particularly in terms of facilitating community self-reliance and local resource mobilization. Yamamoto and Ashizawa (1999, p. 2) confirm this development when they note, 'In many Asian countries, governments have started to reach out to NGOs, viewing them as valuable partners ... The case study of the Panmai Group, for example, points out that Thailand's Eighth National Economic and Social Development Plan (1997-2001) emphasizes the concept of multiparty participation in development efforts, which incorporates the public sector, the private business sector, NGOs, the mass media, academic institutions, and private citizens.'

Interestingly enough, the efforts of business to produce a top-down response to counteract the 'bottom up' grassroots environmentalism of the 1970s and 1980s has affected environmental NGO relationships with the Thai state. In the case of the Environmental Fund, for example, Lee and So (1999) note that the government perceives the fund to be one of the main means to encourage the business sector to develop joint efforts aimed at improving the country's environmental problems. At the same time, however, NGOs applying to the fund for support are often deterred or prevented by the complicated procedures involved in proposing and winning authorization from the fund committee. Since the establishment of the fund almost a decade ago, only sixteen NGOs (as of 1999) had won support for their submissions. This statement of cooperation with limited mechanisms for implementation suggests that the Thai state may well have internally competing visions of how to address environmental issues (that is, with non-corporatist NGOs and/or collabourations with corporate interests) as well as, of course, limited funds with which to do so.

Though much of the discussion in this chapter has focused on Thai NGOs with strong corporatist ties, there are also organizations self-identifying more strongly with anti-development and to some degree anti-state ideologies (see also Farrington et al., 1993). NGOs tend to be associated in the popular imagination of industrialized nations with grassroots action and participatory decision-making. Indeed, the United Nations defines NGO as any 'voluntary, non-profit organization of citizens' (in Ritchey-Vance, 1991, p. 28). However, as argued in this chapter, the Thai state in devising environmental policy strategies is depending increasingly on corporatist non-governmental entities to design and implement programmes that address environmental obstacles to ongoing development. Thus, at least in Bangkok, international donor agencies should be wary of immediately associating NGOs with participatory and grassroots decision-making processes that ultimately lead to enhanced democratization. If international donor agencies are looking solely for more cost-effective mechanisms and institutions than state agencies for public service provision, NGOs may provide a less centralized strategy for designing and implementing programmes. However, NGOs may not provide a significant challenge to social relations that privilege the state, industrial, and military elite in Thailand. We argue instead that the situation in Thailand mirrors Wolch's notion of the shadow state as a para-apparatus of the Thai state, especially given the strong linkages between many non-governmental environmentally-oriented groups, such as Magic Eyes, and Thailand's corporate, political and military elite.

In some ways, an increasing reliance on NGOs, particularly those that are corporatist in nature, represents a devolution of environmental policy responsibility from the centralized Thai state bureaucracy to private sector interests. As in industrialized nations, however, devolution often translates into a deepening of the uneven distribution of resources with low-income tenure insecure communities bearing a disproportionate share of the hazards and the costs (Alexander et al., 1999). Even with these potentials for consolidation of power and hierarchical social relations, the increasing role of NGOs in environmental policy-making in Bangkok also offers the potential for the transfer of influence and power, however slight, from the Thai political and military elite to other stakeholders in the decision-making process, particularly the middle class. Thus, while environmental governance that relies increasingly on NGOs and community participation may not result in short-term realignments of political power, micro-scale and community-level (and centred) actions may provide in the long term opportunities for overt and covert resistance to the *status quo*.

Chapter 6

Lessons for Improved Environmental Management

Environmental management in the cities of both the developed and developing worlds is clearly both an art and a science. It is no longer sufficient, anywhere on earth, to rely only on the efficiency of public works departments to install and maintain expensive physical infrastructure given that even in the wealthiest cities and regions of the world, environmental problems threaten to overwhelm available financial and human resources. It is now generally accepted that, in the future, most urban households will pay a great deal for many traditional public services such as water, sanitation, solid waste collection and disposal and, even, measures to control air pollution. At the same time, the role of urban residents in the design and delivery of services is changing as well. More and more cities look to their residents to actively engage in creative debate and innovative decision-making about how best to provide environmental services. The role of the average resident can be as simple as voting to help select from among a variety of possible service options to as complicated as serving on committees charged with strategizing and implementing various service delivery plans and methodologies. In any case, environmental management must now, more than ever, take into account the values and preferences of the people it is designed to serve.

Lessons from Attitude and Behaviour Research

In the case study portion of this book, we reported on a household survey targeting squatters and slum dwellers in Bangkok that indicated ways in which both cultural and environmental attitudes and behaviours need to be incorporated in the management of environmental infrastructure and basic services. In terms of knowledge and behaviours, for example, we found that slightly less than half of all the participants in the study were aware of community-based organizations in their neighbourhoods, and approximately 30 per cent claimed to have contributed in some way to community-based efforts during the past year (with most activities related to some aspect of community clean-up). Furthermore, most of the individuals surveyed said that they thought almost everyone, including individuals, communities and different levels of government, could work together to improve and protect the quality of community environments.

These generally positive findings for enhancing community participation in environmental management contrasted, however, with several behavioural, attitudinal and cultural findings gleaned from the survey that suggested that it might be more difficult than planners and international donors expect to depend on community participation to lead or facilitate environmental improvements. A clear majority of all respondents, for example, noted that their wastewater was dumped untreated into open sewers and although approximately one-third of the individuals surveyed experienced at least one episode of diarrhoea or gastroenteritis during the previous month, over 50 per cent of the respondents did not believe that there was a relationship between gastroenteritis and water quality. There are obviously still a number of environmental and health knowledge challenges in terms of public understanding regarding the important linkages between environmental conditions and health that must be addressed by policymakers and non-governmental organizations (NGOs). In addition, the survey uncovered an important disconnection between the environmental problems emphasized by the popular press and the BMA budget, in that while many policies and directives focus on traffic, noise and air pollution, the urban poor find the constant and dangerous presence of insects and vermin to be one of their most overwhelming environmental

"Many policies and directives focus on traffic, noise and air pollution, but the urban poor find the presence of insects and vermin to be one of their most overwhelming environmental problems."

Source: Carrie Mitchell (Thailand)

problems. We are almost certain that these same problems, that is, knowledge about the relationship between specific environmental problems and health and the under-investment in vector control from the perspective of low-income residents, are widespread in Southeast Asian cities.

We also used the quantitative data collected in our survey to develop multivariate logistic regression models to further illuminate the linkages among socio-demographic variables, cultural values, attitudes and behaviour. Our exploration of the relationship among several important environmental customs, including individual treatment or purchase of water, sanitation behaviour and wastewater practices, suggested two important and potentially divergent policy implications for improving the environmental and health conditions faced by low-income urban residents. A clear message derived from the quantitative analysis was that level of formal education and household income play a vital role in Bangkok's slums, particularly in terms of relieving environmental distress and in defining the implementation of public policy within the metropolitan area. Most positive environmental actions or behaviours, such as boiling drinking water, placing waste in appropriate bins and visiting doctors when appropriate or necessary, are positively associated with higher income and education levels. Not surprisingly, or uniquely, elevated socio-economic status played a very important role not only in terms of individual health concerns but also in terms of proximate environmental quality. These findings, we believe, are likely to be robust across Bangkok's metropolitan region as well as in most large Southeast Asian cities.

At the same time, working to raise the level of socio-economic status of the urban poor in Bangkok, and in other countries with complex socio-cultural norms and values, may work against the notion or capacity of these same households to engage in community efforts to prevent further local environmental degradation. As noted in Chapter 4, our analysis suggested that longer tenure (in terms of the number of years residents live in a slum or squatter settlement) and a lack of support for the concept of patron-clientism, were both positively associated with the willingness or ability to participate in community-oriented activities and events. As many have noted before, longer tenure in a community is often accompanied by an enhanced sense of belonging and ownership. Hous-ing experts, such as Turner (1969) and Jimenez (1987), for example, have long argued that governments should do all they can to increase the security of tenure available to squatters to provide them with legal incentives to invest in their dwelling quality. It is not surprising to note that a similar calculus is at work in

"Longer tenure was associated with the willingness or ability to participate in community-oriented activities and events."

Source: Kate Swanson (Vietnam)

terms of the relationship between tenure and increased incentives to participate in improving local conditions, as longer-term residence is central to
notions of 'community' in both the industrialized and industrializing world.

A lack of support for the cultural norm we refer to as a belief in patron-client
relationships, however, suggested that there was a more complicated and
problematic relationship that may exist among community participation, socio-
economic status and cultural values. Traditionally, and even today,
patron-client relationships remain one of the primary means that Thai residents
use to help improve or sustain their socio-economic status. A lack of support
for this rather common and well-accepted cultural norm may indicate that
some of these individuals are effectively excluded from the environmental and
household health improvements associated with income and education
discussed previously. At the same time, respondents to the survey suggested
that they were less likely to believe in hierarchical social relationships, and
therefore tended to support engaging in collective behaviour for environmental improvement. Collective behaviour was frequently at odds with the very
'individualistic' Thai society which generally holds that individuals and
families are the basis of all economic and social action. In essence, when
respondents or households believed that they were unlikely or unable to access
traditional patron-client connections (due to their lack of socio-political
networks or their lack of support for non-merit based systems of advance-
ment), they used opportunities for participating more actively or creatively in
community-based efforts benefiting many unrelated households and neighbours. They were more able to perceive or comprehend the benefits that accrue
from collective rather than individual action in that collective action provided
access to material and political benefits for people less connected to patrons
within traditional Thai society.

There is clearly a complex relationship among attitudes, values and
behaviour which implies that simple solutions ignoring these connections may
not be effective in the long or short terms. It might be, as in the case of
Bangkok urban settlements and slums, important to design environmental
policy or programmes to support or facilitate community participation while
simultaneously trying to encourage network building among those who
undermine or are excluded from traditional patron-client relationships. It is
likely that, eventually, strategies supporting community participation will be
vital components or building blocks of the participatory citizen-government
models that many argue should and will replace more conservative and
existing means of urban governance. Furthermore, it is possible that support of
community participation could help supplant the positive connection between
socio-economic status and support for patron-client relationships that we
found in our data. As in most Southeast Asian countries, such an approach is
worth considering in that many parts of the region are following a more

Westernized path to economic growth and development, supplanting many traditional business practices.

Consequently, we argue that it is very important to realize that cultural values and attitudes within Thailand and other countries clearly shape a variety of environmental practices such as community participation, health behaviours, treating water, and disposal of solid waste. Unfortunately, relatively few researchers and scholars have attempted to discern and describe the connections among values, attitudes and behaviours, particularly in the developing world. One clear lesson of the research discussed here is that the changing character of cultural values and attitudes in varying populations responsible for and affected by environmental practices needs to be studied further to understand the significant linkages, and their possible role in policy formation and implementation.

Lessons from Analysis of the NGO Sector

Much of what we learned from the household survey data reinforced the conventional wisdom in that social, economic and political power appeared to be very limited among the urban slum residents. For this reason, the slums of most large Asian cities are hospitable to and attract the attention of various NGOs, some of which seek to improve the environmental quality of local communities (Lee, 1998). Not surprisingly, a number of the lessons regarding community participation and environmental management come from our analysis of the NGO sector in Bangkok.

In Thailand (as in many developing countries), a potential alternative path for those individuals and groups interested in affecting or influencing government policy, particularly if they constitute the educated middle class, is to join or form an NGO. As we note in Chapter 5, as the focus of NGOs in Thailand has shifted during the last decade from general organized resistance to authoritarianism to coalitions with specific and limited concerns, it is difficult to assess to what extent NGOs can actually provide or facilitate broader participation in decision-making or help to create a more responsive and democratic Thai state (Eccleston and Potter, 1996).

As we have explored in detail, the nature of Thai culture and politics generally discourages, if not actually impedes, public participation and collective action. Thai society emphasizes a number of qualities or norms, such as individualism, patron-client relationships and fatalism, which help to create a lack of interest and support for community-based efforts aimed at ameliorating local environmental conditions (Daniere and Takahashi, 1999). The Thai state and its bureaucracy have done much to reinforce the cultural norms within Thai society that emphasize individual (and family) primacy.

The peculiar mix of *laissez-faire* development and policy adopted by the ruling families, as it were, has effectively curbed the efforts of more resistant NGOs. This is particularly true in the sphere of Thai environmental policies at the grassroots level where, during some administrations, the national government actively sought to eliminate the small policy-making role NGOs had managed to carve out for themselves.

At the same time, and this is true of most state bureaucracies in developing countries, the complex and overlapping structure of the Thai government works to obstruct the access of low-income and marginalized populations to the public sector. The confusing and overlapping responsibilities of the myriad of ministries create an intimidating aura around the bureaucracies and this, in itself, is enough to discourage and prevent many disenfranchised groups from even attempting to make demands on the state to protect or rehabilitate deteriorated environments. Not only does the confusing and irrational structure of the bureaucracy hamper civic participation in the public realm but it also makes it very difficult to design and, most importantly, implement public policy programmes, particularly those aimed at improving community environments.

Despite these impediments to organized collective action, even in Thailand, NGOs engaged in community organization and mobilization are often the preferred recipients of international aid. Some international donors are reluctant to rely on what they perceive to be ineffectual state bureaucracies to address such difficult challenges as environmental management in rapidly industrializing and urbanizing regions such as the cities of Southeast Asia. Even the more conservative and traditional donor organizations, such as the World Bank or the International Monetary Fund, are attempting to incorporate community-based initiatives instead of relying exclusively on hierarchical policy-making that has often excluded marginalized individuals and households. As Bebbington (1997) and others have noted, NGOs are viewed to be more pluralistic and bottom-up in nature which allows them, in the optimal conditions, to provide the most disenfranchised of society access to some of the decision-making apparatuses of the state.

Not surprisingly, however, the nature and the effectiveness of NGOs depends, to a great extent, on their relationship to the state in question as well as to many other social and political realities. The degree to which NGOs include all strata of society, and their ability to influence social, economic and environmental policy, clearly differ within and across countries and it is difficult to draw sweeping lessons from individual examples (Farrington, et al., 1993; Levy, 1996). None the less, despite historically antagonistic relationships between non-governmental groups and the Thai bureaucracies, in more recent years environmental NGOs (which are recognized increasingly as non-radical organizations tied to powerful corporations, the government and the military elite) are influencing the decision-making process. Thus, in a trade-off that is undoubtedly mirrored in many parts of the world, NGOs that are confronted with less aggression from the state (read: bureaucracies and politicians) can begin to influence policy, but at the same time must, in order to ward off state aggression, be willing to become more corporatist in nature or, at least, to participate actively with actors from the private sector.

The evidence available on NGOs in Thailand suggests that the relationship between NGOs and the state has been limited by the capacity of NGOs to gain international attention and resources. This limitation, however, has not prevented all NGOs from working with the Thai bureaucracy on particular endeavours, most of which are not of great interest or relevance to the military or corporate elites. The Urban Community Development Office, a Thai state-NGO collaboration, is just one example of a relatively effective effort with limited resources and power that plays a role in coordinating and funding community-based activities. The Thai state has also explicitly stated in a number of plans and legislative reforms its agreement, in principle, to work with NGOs. Since 1999, for example, when the Thai government first created the Environment Fund, the Thai state has assisted to a limited extent with the costs of environmental protection and natural resource conservation projects, most of which were guided and implemented by NGOs. The Thai state has continued to express its willingness to assist with grassroots developments although the funding available for environmental programme has shrunk noticeably in the wake of the re-cent economic crisis. In sum, as is true of other Southeast Asian countries including Vietnam, Indonesia, the Philippines and Malaysia, the Thai state is beginning to give NGOs more institutional space and to allow them a more influential role in the country's political apparatus, particularly when it comes to encouraging community-based initiatives and maximizing local resources.

While many perceive this as a positive development, there are a number of Thai NGOs, as in other places, that reflect more radical ideologies than the NGOs that appear to be gaining influence in the political decision-making. Although much of our analysis focused on the strong corporate connection of many influential Thai non-governmental agencies, it is possible to find groups that see themselves as anti-statist. As we noted earlier, in the eyes of many in the industrialized world, NGOs are conceived of as primarily grassroots, progressive and participatory organizations with little interest in cultivating state or private sector clients or networks. While there are some NGOs in Thailand that fit this more idealized vision of groups with a redistributive and activist agenda, in fact, many tend to be associated with traditional power bases in Thailand.

Thus, the Thai state tends to rely more and more on corporatist non-governmental entities to design and implement programmes that address environmental obstacles to ongoing development. Given this tendency, international donor agencies in Bangkok (and perhaps elsewhere) need to scrutinize their NGO partners more carefully if they are interested in actually promoting participation and grassroots decision-making that many donors believe to be the key to better urban governance. The hope of donors is that the participatory and inclusive process supposedly espoused by NGOs will not only address the specific environmental or infrastructure problem at hand but also eventually even result in increased democratization.

It would be a mistake to conclude this section on the lessons learned from NGOs without noting much of the good that these organizations have achieved despite their problematic relationship with the state and the corporate sector. It is true, for example, that many NGOs can provide an important opportunity to supply a more decentralized means for creating and implementing programmes. While most NGOs may not necessarily also present an immediate challenge to the social relations that exist in a country they, in the long run, can create opportunities among members to initiate change that will result in the inclusion of the previously disenfranchised (Lee, 1998). The learning experience provided by collective action, for example, may do more then simply replicate the shadow state found in Western democracies and begin to create opportunities for individuals to initiate the movement of influence and power from the country's elite to other stakeholders in the decision-making process, particularly the middle class who may, in the end, constitute an effective challenge to the *status quo*.

General Lessons

In addition to the lessons learned from our research on the Thai state and in numerous slum communities of Bangkok, there are a number of other, more general lessons that are very relevant for policy-makers, international donors and communities. One of the important themes gleaned from our research in various countries as well from the case study of Bangkok is that, in most cases, neither private suppliers nor public enterprises or authorities feel legally required to provide basic environmental services such as water, sewage, waste collection or clean air to the bulk of the urban poor. Most contracts between municipal governments and private contractors, as in the provision of piped water in Jakarta for example, contain few if any provisions that require delivering services to informal settlements or squatter areas. Even in the cases where contracts do define such responsibilities, the terms of such contracts often do not consider the possible contributions of other actors, such as the public sector or civil society, that is, NGOs, community groups and associations.

Furthermore, in many if not most instances, while improved service delivery and coverage is clearly a priority with most private and public entities (at least for the middle-class and wealthier neighbourhoods), most providers plan and implement systems that are conventional and technically inappropriate for many low-income communities (Carmen de Mello Lemos, 1998). These systems in turn imply costs unaffordable to low-income groups as well as unaffordable to urban governments that often find that they must unexpectedly subsidize the services if regular and quality service is to be delivered.

At the same time, there are also opportunities to improve the responsiveness of municipal governments to the needs and conditions of the urban poor in Southeast Asian cities. For example, there are a growing number of successful experiences of improvements in low-income settlements implemented and designed by networks comprised of groups and organizations representing a variety of different sectors. The amalgamation of these stakeholders' available means has created situations where all parties involved have improved their lives and met some of their objectives (Anschütz, 1996).

Even in Southeast Asia, where cultural patterns generally discourage public participation and critique, residents of low-income settlements are increasingly reluctant to remain passive in the face of a paternalistic style of governance. Many households and individuals have begun to engage in both the design of systems and in the implementation of projects. Even more remarkable is that they are also willing and able to bear a proportion of the costs of neighbourhood improvement and maintenance. In many slums and squatter communities, there are a range of stakeholders who are willing to undertake participatory activities that can (and probably should) include the provision of basic services. Consequently, the potential has risen for innovative and flexible solutions to ameliorate the provision of services that also help protect the environment. As such, it makes sense for a local or national government to include low-income groups from the beginning of the policy design process in order to benefit from their experience and capabilities. We can even argue that cities or countries that ignore the potential of low-income households to participate in solutions and service provision risk falling behind politically and even economically when compared to countries that are more inclusive and participatory in the way they design and implement environmental services. This is especially true given the stark realities of economic change dictated by ongoing global shifts, where municipal, regional and national governments may be less able to pay for expensive yet necessary improvements to urban infrastructure.

A number of potential strategies for improving the coverage and affordability of public services in poor urban communities has been suggested by a myriad of sources. In the case of contracting with private suppliers, it is very important to note that a concession contract to supply water or waste services in a city where there are urban slums cannot be either socially or economically sustainable if the supplier plans to provide a singly homogeneous service with no variation in levels of service throughout the region. A government that hopes to provide services in a comprehensive fashion to an ever-increasing number of residents living in 'informal' settlements must require that providers offer different levels of service at different prices. For example, some households in some areas should be offered community toilets or public water standpipes or rubbish collection from a community rubbish bin or large, government-provided waste bin rather than individual or house-to-house services.

In addition, concession agreements or the terms of reference for a public enterprise charged with service provision should contain financial incentives for provision of services to low-income slums or settlements. This might include giving a higher weight in terms of performance standards or targets to the number of households provided for or connected within informal settlements. In addition, the government unit might consider subsidizing the private provider or providing extra funds to the public enterprise for each household or community served at adequate levels of quality.

Another important strategy for governments that hope to improve the quality of life and the environment experienced by all urban residents, rich and poor, requires designing projects or public investments to include mechanisms that enable community groups, NGOs and others to contribute the resources and the knowledge they possess. Basically, this means that service provision, what is provided, where, how often, how well and at what price, needs to be negotiated and developed with the input of community members. The right to participate and to help make decisions needs to become as important an aspect of service provision as the user charges assessed to beneficiaries. Governments can do much to design projects or contracts that require community input if desired and can reward both the community and the providing agency directly for their participation, whether through prioritizing the community, subsidizing the provision of services or using their taxation authority, among other available tactics.

The right to contribute and participate can be assured with the assistance of independent external consultants or consumer protection agencies or even ombudspersons selected in consultation with different communities. In addition, the right to arbitration if there is some disagreement over the terms of service delivery might also help avoid the appearance of paternalistic or biased political interests.

While none of these simple strategies to increase and facilitate participation and contributions of the urban poor will necessarily immediately solve the challenges of institutional overlap and lack of technical and human resources that plague the governments of many countries and cities, they are strategies that can result in better service provision and higher quality of life than is currently the case. The irony is that many low-income urban residents spend a great deal of money and time managing and coping with their environmental conditions and there is little recognition of this contribution by either politicians or bureaucrats (Douglass, 1999). As discussed in Chapter 5, some government actors deliberately move to limit and control the communities through their lack of access and opportunity rather than supporting their efforts to reclaim their environments and improve their quality of life.

In terms of designing appropriate and participatory systems, however, it is also important to note that the descriptive analytical work detailed primarily in Chapters 3 and 4 should not be used as a substitute for careful economic and policy analysis of proposals to improve sanitation, water treatment, and solid waste disposal. For example, to be more useful for the actual pricing of services and to correctly estimate their costs and benefits it is important that household surveys of target communities include questions that establish the environmental preferences and requirements of local residents as well as well as individual or household willingness to pay for carefully delineated examples so that a contingent valuation analysis can be conducted. This additional source of information would enhance the attitudinal, behavioural and cultural data that we have presented in this book.

This is not to say, however, that economic analyses are sufficient but only that they are necessary to efficient service delivery. Any economic evaluation that does not contain specific and detailed information about the current circumstances and behaviours of slum residents as well as their attitudes and beliefs may well fail at providing adequate, affordable and appreciated services. It is frequently the case that public services, such as water and sanitation or waste collection and disposal, are provided without consulting or including consumer preferences and perceptions and are met with hostility if not sabotage on the part of the people who the services are supposed to help. Low-income residents, in particular, resent being asked to pay or contribute for services that they do not believe they need or are not designed in a way that they can comfortably access. A project that relies on disenfranchised households to maintain and monitor their implementation may well be doomed to failure. A number of authors have argued that giving local residents input and responsibility, particularly with regard to the implementation and monitoring of projects, can dramatically increase project quality and efficiency because the local residents have such a large stake in improving the quality of services they receive (Tendler, 1997). This argument is clearly supported by much of the evidence we present in this book providing governments and donor agencies with even more incentive to include community groups in the planning and design process from the outset.

Chapter 7

Community Environmental Management in the Developing World

We began this book with the argument that urban environmental conditions have degraded at a rapid rate during the past decade in developing countries along the Pacific Rim. This urban environmental decline during unprecedented economic expansion has led to increasing public health concerns as well as widening challenges to sustainable growth and prosperity. To explain and illustrate this collision of economic expansion and environmental degradation, we outlined the parameters of economic growth and the recent financial crises across Asia and Southeast Asia, and discussed how nation states in the region have sought to manage the negative impacts of economic change (both explosive growth and precipitous decline) as they relate especially to urban environmental conditions. We have argued in this book that appropriate and effective policies to stem this rapid and severe deterioration in urban environmental quality in Bangkok, Thailand and in the developing world more generally, will require a combination of state, non-governmental and community participation.

To explore how state, non-governmental and community actors both exacerbate and ameliorate urban environmental degradation, and to illuminate the interconnection among these actors and the institutions within which they operate, we focused in various chapters on the underlying structure and function of environmental governance in Bangkok. We examined the form and function of the Thai state to clarify the ongoing inability and lack of capacity in Bangkok specifically and Thailand more generally to counteract the negative environmental impacts of rapid economic, social and political change. Perhaps our most important conclusions, however, emphasize the potentially crucial role of the non-governmental sector and local communities in effectively addressing urban environmental issues in Bangkok. We posited that the Thai non-governmental and voluntary sector have an increasingly important role not only in implementing environmental policies at the community, municipal and regional levels, but also in placing environmental issues on the political agenda and designing environmental planning strategies and policies. Beyond the highly visible role of the Thai state, and the increasing engagement in environmental policy of the non-governmental sector, we also underscored the importance of low-income tenure-insecure households in understanding the opportunities and challenges of urban environmental policy and planning in Pacific Rim cities in the developing world.

We used the case study of Bangkok, the capital city of Thailand, and one of the primary cities of the industrializing Pacific Rim, to highlight the complex nature of environmental governance, specifically the interconnections of the state, the non-governmental sector and low-income urban communities. We argued that the Thai state at many levels has both repressed non-governmental environmental activities and also encouraged specific private sector and corporatist forms of non-governmental organizations (NGOs) in dealing with urban environmental conditions. Included in our discussion were an institutional examination of the Bangkok local state, an historical description of Thai NGOs and an analysis of community-level survey data that explored the linkages among Thai culture, household perceptions and environmental practices. From statistical analysis of community-level survey data through to institutional and historical analysis of state and non-state organizations participating in environmental management, we have used distinct lenses with which scholars, policy-makers and communities can understand the components and interconnections characterizing environmental governance in Thailand and Southeast Asia.

Our conceptual approach centred on the interrelationships among the state, the non-governmental sector and local communities. In so doing, we hope to have shed light on the challenges faced by varying stakeholders in dealing with environmental degradation in rapidly changing urban, economic and political contexts. For example, Chapter 5 explained how non-governmental environmental activities in Thailand might be better conceptualized as an instance of an expanding corporatist shadow state, rather than solely or predominantly an illustration of community grassroots mobilization. In situations where international donors and government agencies seek to empower and engage low-income urban residents, or to include community views in environmental policy design and implementation, the definition of non-governmental activity as constituting a corporatist shadow state should temper overexuberant enthusiasm about the potential democratizing and decentralizing role often attributed to non-governmental involvement. Indeed, we have sought to highlight the complexity and difficulty in ensuring community participation in environmental management given the structure and interrelationships of the Thai state, increasingly corporatist non-governmental organizations in Bangkok, and the challenges faced by low-income residents and communities.

The tensions inherent in environmental governance in rapidly changing urban areas will undoubtedly require difficult decisions concerning trade-offs to be made in improving urban environmental quality during rapid economic shifts. Efforts to improve urban environmental quality during the stark economic decline across Asia and Southeast Asia in the late 1990s provides a good example of such challenges. With depleting state reserves and unsustainable economic growth, non-governmental and community participation, with all the political and social complications inherent in such multiple interests, will probably become an increasingly common element characterizing environmental governance in the Pacific Rim.

In explaining environmental governance as a process and outcome inherently embedded in social relations that centre on the state, the shadow state and urban dwellers, we have sought to steer away from any single conceptualization for analysing and promoting community participation in environmental management and policy. A single approach to community participation would be likely to provide a clearer mandate for non-governmental expansion and resident involvement in environmental policy design and implementation. Explanations based on community attitudes, and one that we used in Chapter 4, often begin with the premiss that top-down, centralized environmental policies are often not appropriate for local communities, therefore, such policies are likely to be ineffective. Instead, a focus on community beliefs highlights the view that urban residents, especially in Bangkok, may have attitude systems contrary to communal goals that enhance urban environmental conditions, and that with appropriate information and incentives, urban residents will engage more frequently in activities and behaviours that promote environmental quality and, consequently, enhance public health and economic growth. When attitudes are deemed to be the ultimate source of a lack of communal engagement in environmental management, information, education and community input become the remedies for rapid and severe urban environmental deterioration. We showed in Chapter 4 that this approach can be quite useful in identifying the clear disconnections between Thai state environmental management policies and community-level practices, and indicating where, when, and how the Thai state (at various levels) and NGOs might work with community residents to improve urban environmental conditions.

However, we have also argued, particularly in Chapters 5 and 6, that while this perspective indicates a relatively clear set of strategies and rationale, it also tends to overlook the complexity and constraints involved in environmental management among individuals and institutions. Among individuals, as we argued in Chapter 6, attitudes and values, while crucial for identifying the broad parameters of infrastructure needs and obstacles, are not detailed enough to indicate the specific types of policies and programmes necessary to respond to differential geographies, local economies and community environmental situations. More broadly, however, such a perspective does little to elucidate the underlying relationships and challenges embodied in the structure and actions of institutions such as the Thai state and corporatist non-governmental entities that may preclude significant institutional change for environmental policy improvements. This implies, then, that even with a focus on enhancing community participation, especially by low-income tenure insecure residents, solutions drawn solely from community attitude studies are likely to be partial and potentially repressive in specific circumstances.

Rather than pursuing an all-encompassing framework, we have used a broad emphasis on the state, non-governmental sector and communities to illustrate how institutions have varied in their approach to each other and to environmental policies through time and across geographic locations within the city of Bangkok. In this chapter, we shall continue this line of thinking, that effective local environmental management cannot only be seen as the responsibility of the state or local communities. Indeed, although the inclusion of communities, especially those that are low-income and tenure insecure, will broaden the dialogue concerning specific urban environmental management issues and policies, what it may not do is to guarantee short- and long-term improvements in environmental quality. We are interested not only in improving environmental conditions in the short term, but also in highlighting the groups and mechanisms that will enhance environmental quality especially for low-income tenure-insecure residents in Bangkok in the long term. In this sense, we echo the call for a 'local scale version [of urban sustainability] in which sustainability is synonymous with *sustainable livelihoods* and in which local context can lead to different and locally contingent perspectives on the meaning of and conditions for sustainability and the means to achieve it' (NSF Workshop on Urban Sustainability, 2000, p. 6). We will suggest in this concluding chapter that an empowering community participation in environmental management in Bangkok (and the developing world more generally) should be understood as intricately intertwined with three concomitant factors: the ongoing urbanization of Pacific Rim cities, and the uncertainty and instability of local and regional economies; the short-term intractability of environmental degradation; and the need for community ties, which some have termed social capital, to catalyse mobilization for long-term community participation.

Obstacles to Community Environmental Management

The rapid economic expansion in the ASEAN region during the early 1990s brought uneven wealth to the populace of member nations, and urban environmental deterioration in these nations' major urban areas. The economic downturn in the late 1990s exacerbated the negative impacts of rapid growth, including expanding residential concentrations of tenure-insecure and low-income households, continued environmental quality decline, and now unstable and uncertain local and regional economies. Along with this urban environmental degradation and slowdown in economic expansion, there has been increasing political instability, widening violence and expanding poverty. There is little doubt that political instability and intense poverty have continued during the early 2000s across the Southeast Asia region, with rising political protests, calls for new leadership and violence in Thailand, Indonesia and the Philippines. The combinations of economic decline and environmental deterioration across Southeast Asia were discussed in Chapter 1.

In chapters throughout the book, three potential societal and institutional sites for enhancing environmental governance were analysed. First, as discussed in Chapter 3, the Thai state at various levels has been the primary provider of infrastructure services in Bangkok and, therefore, has had the longest and potentially deepest experience with dealing with environmental problems in the Bangkok Metropolitan Region. The second institutional site, discussed in Chapter 5, and in Bangkok largely related to the Thai state, consists of the highly variable non-governmental sector in Thailand, much of which has become corporatist in nature, especially with respect to improving urban environmental conditions in the BMR. Although there have been NGOs that focus on community empowerment and resistance to the Thai state, contemporary NGOs devising highly visible programmes geared toward environmental quality improvements have been largely corporatist and intricately linked to powerful political, business and military interests. The third site for environmental governance, discussed in Chapter 4, is less institutional and more *ad hoc* in Bangkok, and consists of community participation at the micro, meso and macro levels. Community participation has been touted by international donors and increasingly by governments as a panacea for implementing urban environmental policies; the incompatibility of top-down infrastructure strategies with household behaviour has been argued to be the source of policy ineffectiveness. These chapters have elucidated the complexity underlying each of these varying societal sites for environmental governance, not only pointing to the difficulties in designing and implementing effective urban environmental policies, but also highlighting the broader issues of poverty management and social control through environmental management and non-governmental activity.

There are obviously no simple solutions to urban environmental degradation in Bangkok. One of the reasons for this is the short-term intractability of deteriorating urban environmental conditions. Even in good economic times, effective environmental management will require a combination of micro-level strategies (for example, throwing solid waste in available bins), meso-level strategies (for example, communal clean-up and potable water provision to expanding settlements), and macro-level strategies (for example, regional wastewater treatment and the development of environmentally sensitive solid waste disposal methods). Especially in the face of economic decline, urban environmental quality is likely to take a back seat to the need for economic expansion. Short-term intractability contributes to public dissatisfaction with public sector efforts to cope with declining urban environmental quality of life, and has led to a focus among Thai voters to revive the failing economy at any cost.

Throughout the book, we have emphasized that along with the Thai state and the increasingly corporatist non-governmental sector, communities and settlements represent a vital and largely untapped resource for environmental governance. Although urban settlements are often viewed by policy-makers and scholars as problematic, they can also be characterized as embodied with energy, collective concern and creative potential for problem-solving. As alluded to in previous chapters, a lack of focus on community-level concerns, especially in informal urban settlements, has reinforced the prevailing view by the state and perhaps NGOs that such settlements embody the worst of Thai society – a focus on individualism at the cost of communal quality of life, a lack of understanding and/or sensitivity about the broader impacts of individual behaviour and the inability to become an important and powerful entity in Thai political and economic life. Because a lack of tenure security and low incomes are also largely linked to a lack of connection with powerful patrons and influential networks of military, political or business elites, environmental policy and management are also largely confined to the realms of the Thai state and corporatist NGOs.

The discussion in varying chapters of the roles these actors play in environmental governance in Bangkok highlights the deep social divisions generated by these characterizations of informal settlements that are reinforced by ongoing institutional practices. While households and communities that have potable drinking water, solid waste collection and other infrastructure services have much to gain by maintaining the public sector focus on wealthy and politically influential groups and neighbourhoods, the lack of capacity of public sector agencies, non-governmental and grassroots organizations, and low-income residents to counter the power of these traditional and institutionalized social relations has created a extremely uneven terrain of environmental quality and infrastructure provision. The increasing focus of international donor agencies and governmental bodies on NGOs and community participation has contributed perhaps to the creation of a widening burden placed on low-income tenure-insecure residents, households and settlements to improve their community environmental conditions. That is, in addition to the burdens of meeting basic needs for everyday survival, residents and communities are increasingly being asked to design and implement environmental management policies as volunteers and without remuneration. As this environmental management burden increases for the lowest-income groups in urban metropolitan areas, the responsibility for policy ineffectiveness also shifts without concomitant financial and institutional resources. Devolution thus brings with it potential opportunities for political enfranchisement, but more probably, for these low-income tenure insecure residents and communities, a shifting of the responsibility for urban environmental quality also displaces the blame for ineffective policy design and implementation to the least powerful in Thai society. NGOs, especially those with corporate and political elite ties, are likely to be the biggest beneficiary of environmental policy devolution because of increasing funding support by international donors and focus on community residents to design, implement and enforce such non-governmental environmental policies and programmes.

To challenge the oppressive potential of environmental policy devolution, there is an urgent need for strengthening community ties and promoting grassroots mobilization especially among low-income tenure-insecure urban residents. In Bangkok, the ongoing presence of the Thai cultural norm structure of individualism and patron-client relationships (which would seem to mesh well with corporatist shadow state environmental management strategies) means that grassroots mobilization for communal environmental quality will probably be difficult to initiate and maintain. However, truly participatory environmental management and governance may be facilitated if attention is paid not only to the technical aspects of environmental policy-making, but also to the political dimensions of grassroots mobilization.

What about Social Capital?[1]

Local mobilization is very dependent on the interconnectedness of residents in specific geographic locales. The social networks present in low-income tenure-insecure urban settlements represent the systems of social relations existing among and within households. These systems are concretely expressed as the reciprocal (perhaps asymmetrical) exchange of material goods, services and money as well as less tangible resources, such as mutual assistance and emotional support. In general, scholars have focused on the two types of networks representing contract and reciprocal relations (for example, Granovetter, 1985). The first concerns the social relations of residents within a specific community or geographic locale while the second is determined by the nature of linkages between the community and institutions, such as NGOs, politicians and government agencies, based outside the community but within the urban or metropolitan area (for example, Jessop, 1998; Kim et al., 1997; Takahashi and Rodriguez, 2002).

---

[1] This section borrows heavily from Daniere, Takahashi and NaRanong (2002) and is included with the permission of Edward Elgar Publications.

The potential effectiveness of social networks lies in the social capital they may provide (Coleman, 1988). Social capital, according to Gertler (2002), is comprised of 'those characteristics of social structure or social relations that facilitate collabourative action, and as a result, enhance economic performance'. Although theorized and discussed by many (for example, Coleman, 1988, 1990; Edwards and Foley, 1997; Erickson, 1998; Minkoff, 1997; Woolcock, 1998), Greico's (1995) discussion on the usefulness of social relations is quite useful as a starting point for articulating further research questions on understanding the nature and impact of social capital. Greico defines social capital as those social relations that represent resources which individuals, households and larger social categories mobilize to compensate for situations occasioned by the lack of sufficient income. Social capital then is a potential resource for impoverished communities if residents are able to work together and create assets based on their systems of social relations.

Among these potential resources are exchange of information, credit supply and pooling of monetary resources (Davern, 1997; Granovetter, 1973, 1995; Mitchell, 1995). Such financial social capital can be of use to communities attempting to manage or improve their access to environmental services and consequently may directly affect urban environmental governance. The potential to pool financial resources, for example, might allow a community to invest in an improved drainage system that would not be affordable nor effective on a household basis. Or, formal networks between a community and political agents might allow that community to make effective demands for improved drainage or more regular pick-up of solid waste that would be quite difficult for a community without an established link to a politician.

Early studies on social networks (Granovetter, 1973, 1995) indicate that the strength of interpersonal ties in these networks plays a crucial role in the ability of social network members to promote a collective cause. Weak ties, it has been argued, 'are more likely to link members of *different* groups than are strong ones, which tend to be concentrated within particular groups' (Granovetter, 1973). More recent work on the expansion of informal activity in the face of economic recession further shows that close-knit, ethnic and religious social networks promote particularistic solutions and weaken more broadly-based social organizations (Majumdar, 1995; Meagher, 1995). Strength of ties is determined by the amount of time network members spend with one another as well as the emotional intensity and intimacy involved in such relationships (Granovetter, 1973). To understand more fully how social networks may influence the capacity of low-income tenure-insecure residents to pool monetary and political resources for improved environmental quality, researchers should work to describe and elabourate the nature of social networks and describe how these networks among residents, agencies and institutions influence the capacity of a community to manage local environments. A number of scholars argue that networks among individuals, communities, NGOs and government agencies can either enhance or inhibit the potential accomplishments of a group or organization (Portes and Landolt, 1996; Woolcock, 1998). Therefore, researchers should also study the positive and negative consequences of networking and the use of social capital to improve environmental conditions and infrastructure.

Additionally, the literature suggests that the efficacy of social networks and social capital is widely influenced by government policies that can encourage or discourage the formation of local organizations (Cooke and Morgan, 1993; Evans, 1997; Gertler, 1997; Morgan, 1997; Putnam, 1993). As Chapter 5 indicated, the Thai state has acted in the past and present both to repress and to encourage non-governmental activity.

The state and non-governmental sectors may work to create conditions where grassroots mobilization can flourish. Indeed, analysts of the so-called informal sector, which may include NGOs but also unregulated business and labour market activity, contend that 'the conditions necessary for its co-ordination and operation on a major scale, even in a situation of economic crisis, can only exist in an environment of state complicity' (Meagher, 1995). Likewise, as argued in Chapter 4, government and non-governmental efforts at environmental policy-making are likely to be much more effective given a sensitivity to community-level beliefs and network linkages. A number of researchers have found that businesses in a specific region are more likely to flourish if they incorporate existing social networks and local culture into their day-to-day operating procedures (Storper, 1997). Similarly, we believe that the most successful government policies regarding environmental protection at the local level are those that can be easily incorporated into the social networks and daily routines of communities and that are appropriate to the beliefs and practices of individual residents (see also Lee, 1998). There is a significant gap in our understanding of how state policies and programs relate to and interact with community-level social networks in the context of urban environmental management in developing country cities. We believe, therefore, that researchers should work to understand the means through which overlapping or non-overlapping networks of residents and institutions, and the social capital created through these networks, can promote urban governance related to environmental management. Such work would clarify the relationship between state policy and the form and effectiveness of local social networks.

Many researchers focus on the role and importance of social networks and social capital in the economic or business sphere. Relatively few scholars, however, have made the connection among social networks, social capital and environmental governance. An exception to this is Douglass et al. (2002) who analyse social capital formation within the context of state-community interaction in low-income neighbourhoods in Asia. They find that individual communities in different countries achieve partial solutions to their environmental problems and that the nature of the solutions is related both to macro-level policies as well as the political regime. This means that the local scale in geographic terms, is extremely important to understanding the possible effectiveness of community networks and social capital in terms of environmental management and improved governance. Researchers, international donors and policy-makers should refocus their efforts on understanding these linkages to realize the ideal of effective community participation.

Conclusion

While we have argued throughout this book that environmental management in the Pacific Rim should be understood as the expression of state, non-governmental and resident interrelationships, we have also suggested in this chapter that the contemporary problematic of urban environmental issues is significantly defined by three interrelated factors: rapid urbanization of Pacific Rim cities, and the recent uncertainty and instability of local and regional economies; the short-term intractability of urban environmental deterioration; and the need for grassroots mobilization to counter the potentially oppressive and social-control dimensions of rising non-governmental activity in urban environmental management. Our argument that urban environmental manage-

"Calls for community participation are in response to the unevenness of public service provision across social class and residential location."

Source: http://images.umdl.   *Source*: Molly Davidson-Welling (Vietnam)
umich.edu/i/indonhesian/
(Surakarta, Indonesia)

ment in Thailand and other Pacific Rim developing cities is defined and constrained by these three factors rests on the core assumption that community participation, in and of itself, may not result in either effective environmental policies or an emancipatory or empowering decision-making process. The calls for community participation are based on the twin notions that top-down decision-making results in ineffective infrastructure outcomes and that decentralization is preferable to centralized policy design and implementation. Such calls are in direct response to the inability of centralized governments to cope with rapid urbanization and especially the unevenness of public service provision across social class and residential location. While we can hope that environmental conditions will improve, especially for low-income and tenure-insecure residents in Bangkok and other major metropolitan areas along the Pacific Rim, ongoing political, economic and social tensions and instability in ASEAN member nations puts urban environmental management at a relatively low political priority.

One longer-term means of enhancing community participation in environmental management is to think more broadly about the context within which urban environmental deterioration exists. Poverty, social cleavages and lack of access to decision-making are fundamentally related to the uneven geography of urban environmental degradation in Bangkok. While this may appear at first to be an obvious characterization of environmental challenges in Bangkok and other Pacific Rim cities, the linkages to broader social relations and political economy must continue to be emphasized if longer-term solutions are to be devised. Though immediate results are less likely with this longer-term view, such explanations and action are necessary if international donors, government agencies, NGOs, community groups and residents are to make substantial and long-lasting changes to community environmental quality.

Bibliography

AAMA (Association of Automobile Manufacturers of America) (1998). *World Motor Vehicle Data*. Detroit: Association of Automobile Manufacturers of America.

Ajzen, Icek and Martin Fishbein (1980). *Understanding Attitudes and Predicting Social Behavior*. Englewood Cliffs, NJ: Prentice Hall.

Alexander, Jennifer, Renee Nank and Camilla Stivers (1999) 'Implications of Welfare Reform: Do Nonprofit Survival Strategies Threaten Civil Society'. *Nonprofit and Voluntary Sector Quarterly* **28** (4), pp. 452–75.

Andrews, Clinton J. (1999). Putting Industrial Ecology into Place: Evolving Roles for Planners. *Journal of the American Planning Association* **65** (4), pp. 364–75.

Anschütz, Justine (1996). *Community-Based Solid Water Management and Water Supply Projects: Problems and Solutions Compared*. Gouda, the Netherlands: WASTE/Urban Waste Expertise Programme.

Asia Development Bank (1991). *Country Reports*. Manila, pp. 127–45.

Assad, Ragui (1996). 'Formalizing the Informal? The Transformation of Cairo's Refuse Collection System'. *Journal of Planning Education and Research* **16** (2), pp. 115–26.

Bartholomew, Keith (1999). 'The Evolution of American Nongovernmental Land Use Organizations'. *Journal of the American Planning Association* **65** (4), pp. 357–63.

Bartone, Carl et al. (1994). 'Toward Environmental Strategies for Cities: Policy Considerations for Urban Environmental Management in Developing Countries'. *Urban Management Programme Policy Papers* **18**. Washington, DC: World Bank.

Baud, I., S. Grafakos, M. Hardijk and J. Post. (2001). 'Quality of Life and Alliances in Solid Waste Management: Contributions to Urban Sustainable Development'. *Cities* **18** (1), pp. 3–12.

Baumgartner, F.R. and B.D. Jones (1993). *Agendas and Instability in American Politics*. Chicago, IL: University of Chicago Press.

Bebbington, Anthony (1997). 'Social Capital and Rural Intensification: Local Organizations and Islands of Sustainability in the Rural Andes'. *Geographical Journal* **163** (2), pp. 189–99.

Bebbington, Anthony (2000). 'Reencountering Development: Livelihood Transitions and Place Transformations in the Andes'. *Annals of the Association of American Geographers* **90** (3), pp. 495–520.

Bell, David (1997). 'A Review of Interpreting Development: Capitalism, Democracy and the Middle Class in Thailand'. *Crossroads* **11**, pp. 135–8.

Bello, Walden (1993). 'Trouble in Paradise: The Tension of Economic Integration in the Asia-Pacific'. *World Policy Journal* **10** (2), pp. 33–40.

Bellamy, Carol (2001). *The State of the World's Children 2001: Early Childhood*. New York: United Nations Children's Fund (UNICEF), 116 pp.

Berner, E. (1997). 'Opportunities and Insecurities: Globalization, Localities and the Struggle for Urban Land in Manila'. *European Journal of Development Research* **9** (1), pp. 167–82.

Berry, B. (1982). 'Clean Water for All: Equity-Based Urban Water Supply Alternatives for Indonesia's Cities'. *Urban Geography* **3**, pp. 281–99.

Bomba, W.G. (1982). 'Sanitation Aspects of Water Supply and Excreta Disposal'. In J. Schiller and R. L. Droset, eds. *Water Supply and Sanitation in Developing Countries*. Ann Arbor, MI: Ann Arbor Science Publishers, pp. 223–74.

Brennan, Ellen M. (1995) 'Developing Management Responses for Mega-urban Regions'. In T.G. McGee and Ira Robinson, eds. *The Mega-Urban Regions of South-East Asia*. Vancouver: UBC Press, pp. 242–65.

Briscoe, John (1999). 'The Changing Face of Water Infrastructure Financing in Developing Countries'. *International Journal of Water Resources Development* **15**, pp. 301–8.

Briscoe J., P. Furtado de Castro, C. Griffin, J. North and O. Olsen (1990). 'Toward Equitable and Sustainable Rural Water Supplies: A Contingent Valuation Study in Brazil'. *World Bank Economic Review* **4**, pp. 115–34.

Brooks, Douglas H. (1998). 'Challenges for Asia's Trade and Environment'. Manila: Asian Development Bank, Economic and Development Resource Center.

Bryant, Bunyan (1997). 'Environmental Justice, Consumption and Hazardous Waste Within People of Color Communities in the United States and Developing Countries'. *International Journal of Contemporary Sociology* **34**, pp. 159–74.

Bryant, Raymond L. and Geoff A. Wilson (1998). 'Rethinking Environmental Management'. *Progress in Human Geography* **22** (3), pp. 321–43.

Bunbongkarn, S. (1987). 'Political Institutions and Processes'. In S. Xuto, ed. *Government and Politics of Thailand*. Singapore: Oxford University Press, pp. 41–74.

Cairncorss, Sandy (1990). 'Water Supply and the Urban Poor'. In Sandy Cairncross, Jorge Hardoy and David Satterthwaite, eds *The Poor Die Young: Housing and Health in Third World Cities*. London: Earthscan Publishers, pp. 73–106.

Callahan, William A. (1998). 'Imagining Democracy: Reading the Events of May in Thailand'. Singapore and London: Institute of Southeast Asian Studies.

Carmen de Mello Lemos, Maria (1998). 'The Politics of Pollution Control in Brazil: State Actors and Social Movements Cleaning up Cubato'. *World Development* **26** (1), pp. 75–87.

Choe, Kyeonge, Dale Whittington and Donald T. Lauria (1996). 'The Economic Benefits of Surface Water Quality Improvements in Developing Countries: A Case Study of Davao, Philippines'. *Land Economics* **72**, pp. 519–37.

Clarke, Gerard (1998). *The Politics of NGOs in South-East Asia*. New York: Routledge.

Coleman, J.S. (1988). 'Social Capital in the Creation of Human Capital'. *American Journal of Sociology* **94** (supplement), pp. S95–S120.

Coleman, J.S. (1990). *Social Capital in Foundations of Social Theory*. Cambridge, MA: Belknap Press of Harvard University Press, pp. 300–325.

Cooke, P. and K. Morgan (1993). 'The Network Paradigm: New Departures in Corporate and Regional Development'. *Environment and Planning D: Society and Space* **11** (5), pp. 543– 64.

Corrigan, P. and D. Sayer (1985). *The Great Arch: English State Formation as Cultural Revolution*. London: Basil Blackwell.

Costen, Jennifer M. (1998). 'A Model and Typology of Government-NGO Relationships'. *Nonprofit and Voluntary Sector Quarterly* **27** (3), pp. 358–82.

Cox, K. (1993). 'The Local and the Global in the New Urban Politics: A Critical View'. *Environment and Planning D: Society and Space* **11**, pp. 433–48.

Cox, K.R. and Andrew E.G. Jonas (1993). 'Urban Development, Collective Consumption and the Politics of Metropolitan Fragmentation'. *Political Geography* **12** (1), pp. 8–37.

Cox, K.R. and A.J. Mair (1988). 'Locality and Community in the Politics of Local Economic Development'. *Annals of the Association of American Geographers* **78**, pp. 307–25.

Crane, Randall (1994). 'Water Markets, Market Reform, and the Urban Poor: Results from Jakarta, Indonesia'. *World Development* **22** (11), pp. 71–83.

Crane, Randall and Amrita Daniere (1996). 'Measuring Access to Basic Services in Global Cities: Descriptive and Behavioral Approaches'. *Journal of the American Planning Association* **62**, pp. 203–21.

Crane, Randall, Amrita Daniere and Stacy Harwood (1997). 'The Contribution of Environmental Amenities to Housing Value in Slums: Implications for Cost Recovery in Jakarta and Bangkok'. *Urban Studies* **34**, pp. 1495–1512.

Cumper, G.E. (1984). *Determinants of Health Levels in Developing Countries*. Letchworth: Research Studies Press.

Daniere, Amrita (1995).' Transportation Planning and implementation in cities of the third world: the case of Bangkok'. *Environment and Planning C: Government and Policy* **13**, pp. 25–45.

Daniere, Amrita (1996). 'Growth, Inequality and Poverty in South-east Asia: The Case of Thailand'. *Third World Planning Review* **18**, pp. 373–96.

Daniere, Amrita and Lois M. Takahashi (1997). 'Environmental Policy in Thailand: Values, Attitudes, and Behaviour among the Slum Dwellers of Bangkok'. *Environment and Planning C* **15**, pp. 305–27.

Daniere, Amrita and Lois M. Takahashi (1999a). 'Environmental Behavior in Bangkok, Thailand: A Portrait of Attitudes, Values, and Practices'. *Economic Development and Cultural Change* **47** (3), pp. 525–57.

Daniere, Amrita and Lois M. Takahashi (1999b). 'Public Policy and Human Dignity in Thailand: Environmental Policies and Human Values in Bangkok'. *Policy Sciences* **32** pp. 247–268.

Daniere, Amrita, Lois M. Takahashi and Anchana NaRanong (2002). 'The role of social capital in the environmental management in Southeast Asian cities'. In J. Isham, S. Ramaswamy and T. Kelly, eds. *Social Capital and Well Being in Developing Economies*. Gloucester, UK. Edward Elgar Publishers.

Davern, M. (1997). 'Social Networks and Economic Sociology: A Proposed Agenda for a more Complete Social Science'. *American Journal of Economics and Sociology* **56** (3), pp. 287–302.

Dixon, Chris (1991). *Southeast Asia in the World-Economy*. Cambridge: Cambridge University Press.

Dixon, Chris and David Drakakis-Smith (1993). *Economic and Social Development in Pacific Asia*. London: Routledge.

Dixon, Chris and David Drakakis-Smith, eds. (1997). *Uneven Development in Southeast Asia*. VT: Ashgate Publishing.

Dobbin, Frank and John R. Sutton (1998). 'The Strength of a Weak State: The Rights Revolution and the Rise of Human Resources Management Divisions'. *American Journal of Sociology* **104** (2), pp. 441–54.

Douglass, Mike (1999). 'Geographies of Resilience: Slums, Squatters and Community-State Relations in Seoul and Bangkok'. *Plurimondi* **2** (July–Dec), pp. 213–33.

Douglass, Mike, Orathai Ard-am and Ik Ki Kim (2002). 'Urban Poverty in Thailand: Values, Attitudes and Bahaviour Among the Slum-Dwellers of Bangkok'. *Environment and Planning C: Government Policy* **15**, pp. 305–27.

Dowall, D.E. (1992). 'A Second Look at the Bangkok Land and Housing Market'. *Urban Studies* **29**, pp. 25–37.

Dua, Andre and Daniel C. Esty (1997). 'Sustaining the Asia Pacific Miracle: Environmental Pollution and Economic Integration'. Washington, DC: Institute for International Economics.

Duesenberry, James S. (1967). *Income, Saving and the Theory of Consumer Behavior*. New York: Oxford University Press.

Dwyer, D. (1968). 'The City in the Developing World and the Example of Southeast Asia'. *Geography* **35**, pp. 353–69.

Eade, Deborah, ed. (2000). *Development, NGOS, and Civil Society: Selected Essays from Development in Practice*, introduced by Jenny Pearce. Oxford: Oxfam.

Eccleston, Bernard and David Potter (1996). 'Environmental NGOs and Different Political Contexts in South-East Asia'. In Michael J.G. Parnwall and Raymond Bryant, eds. *Environmental Change in South-East Asia*. New York: Routledge, pp. 49–66.

The Economist (US) (2000). 'Home on a Mountain of Rubbish (Rubbish Dumps are Symbol of Poverty in Philippines)'. 15 July 15 v356 (i8179), p. 41.

The Economist (1996). 'Back on the Road'. 11 May 11, pp. 1–20.

Edwards, B. and W.M. Foley (1997). 'Social Capital and the Political Economy of our Discontent'. *American Behavioral Scientist* **40** (5), pp. 669–78.

Eisinger, Peter (1988). *The Rise of the Entrepreneurial State*. Madison, WI: University of Wisconsin Press.

Elkin, Stephen (1987). *City and Regime in the American Republic*. Chicago, IL: University of Chicago Press.

Erickson, B.H. (1998). 'Social Capital and its Profits, Local and Global'. Paper presented at the Sunbelt XVIII and 5th European International Conference on Social Networks, Sitges, Spain, 27–31 May.

Evans, P. (1997). *State Society Synergy: Government and Social Capital in Development*. Berkeley, CA: International and Area Studies, University of California at Berkeley.

FAO (Food and Agriculture Organization) (1998). 'Forest Resources Assessment 1998: Tropical Countries'. FAO Forestry Paper 212. Rome: Food and Agriculture Organization.

Farrington, John, Anthony Bebbington, Kate Wellard and David J. Lewis (1993). *Reluctant Partners?: Non-governmental Organizations, the State and Sustainable Agricultural Development*. London: Routledge.

Freeman, Donald B. (1996). 'Doi Moi Policy and the Small-Enterprise Boom in Ho Chi Minh City, Vietnam'. *The Geographical Review* **86** (2), pp. 175–197.

Furedy, C. (1997). 'Socio-environmental Initiatives in Solid Waste Management in Southern Cities: Developing International Comparisons'. *Journal of Public Health* **27** (2), pp. 142–56.

Gertler. M.S. (1997). 'The Invention of Regional Culture'. In R. Lee and J. Wills, eds. *Geographies of Economies*. London: Edward Arnold, pp. 47–58.

Gertler, M.S. (2002). 'Social Capital'. In R. Johnston, D. Gregory, G. Pratt, D. Smith and M. Watts, eds. *The Dictionary of Human Geography* (4th edition) Cambridge, MA: Blackwell Publishers, pp. 746–747.

Gibbs, David and Andrew E.G. Jonas (2000). 'Governance and Regulation in Local Environmental Policy: The Utility of a Regime Approach'. *Geoforum* **31**, pp. 299–313.

Girling, J.L.S. (1981). *Thailand: Society and Politics*. Ithaca, NY: Cornell University Press.

Goss, J.D. *(1997)*. 'Right to the City: Forms of Land Allocation and Struggle for Living Place and Working Space in Manila'. *Philippine Sociological Review* **46** (3), pp. 312–41.

Granovetter, M.S. (1973). 'The Strength of Weak Ties'. *American Journal of Sociology* **78** (6), pp. 1360–80.

Granovetter, M.S. (1985). 'Economic Action and Social Structure: The Problem of Embeddedness'. *American Journal of Sociology* **91** (2), pp. 481–510.

Granovetter, M.S. (1995). *Getting a Job: A Study of Contacts and Careers*. Cambridge, MA: Harvard University Press.

Grieco, M. (1995). 'Time Pressures and Low-income Families: The Implications for "Social" Transport Policy in Europe'. *Community Development Journal* **30** (4), pp. 347–63.

Harpham, T.P. Garner and C. Surjadi (1990). 'Planning for Child Health in a Poor Urban Environment – the Case of Jakarta, Indonesia'. *Environment and Urbanization* **2** (2), pp. 77–82.

Hewison, Kevin, Richard Robison and Garry Rodan, eds (1993). *Southeast Asia in the 1990s: Authoritarianism, Democracy and Capitalism*. St Leonards, NSW: Allen and Unwin.

Hiebert, Murray (1996). *Chasing the Tigers: A Portrait of the New Vietnam*. New York: Kodansha International.

Hiep, Nguyen Duc (1996). 'Some Aspects of Air Quality in Ho Chi Minh City, Vietnam'. Paper presented at the Seminar of Environment and Development in Vietnam, National Centre for Development Studies, Australia National University, Sydney.

Higgott, Richard A., Geoffrey R.D. Underhill, and Andreas Bieler, eds (2000). *Non-state Actors and Authority in the Global System*. London and New York: Routledge.

Hirsch, Philip (1994). 'Where are the Roots of Thai Environmentalism?'. *TEI Quarterly Environment Journal* **2** (2), pp. 5–15.

Hoggett, P. (1991). 'A New Management in the Public Sector?'. *Policy and Politics* **19** (4), pp. 243–56.

Hogrewe, W., S.D. Joyce and E.A. Perez (1993). 'The Unique Challenges of Improving Peri-urban Sanitation', WASH Technical Report No. 86. Washington, DC: Water and Sanitation for Health Project.

Hugo, G.J. (1997). 'Population Change and Development in Indonesia'. In R.F. Watters and T.G. McGee, eds. *Asia Pacific: New Geographies of the Pacific Rim*. London: Hurst and Company, pp. 223–49.

Huntington, Samuel P. (1972). *The Soldier and the State: theory and politics of civil-military relations*. Cambridge: Belknap Press of Harvard University.

IIED (International Institute for Environment and Development) (2001) *Environmental Problems in an Urbanizing World*. London.

Isham, J. and S. Kähkönen (2000). 'What Determines the Effectiveness of Community-based Water Projects? Evidence from Central Java, Indonesia on Demand Responsiveness, Service Rules, and Social Capital'. Paper presented at the 42nd Annual Conference of the Association of Collegiate Schools of Planning, Atlanta, GA.

Isham, J., S. Ramaswamy and T. Kelly (eds) (2002). *Social Capital and Well-Being in Developing Economies*. Glouchester, UK: Edward Elgar Publications.

IMF (International Monetary Fund) (2001). 'Thailand: Selected Issues'. Country Report No 01/147.

Jacobi, P. (1990). 'Habitat and Health in the Municipality'. *Environment and Urbanization* 2 (2), pp. 33–45.

Jacobs, Norman (1971). *Modernization without Development: Thailand as an Asian Case Study*. New York: Praeger Publishers.

Jellinek, Lea (1991). *The Wheel of Fortune: The History of a Poor Community in Jakarta*. Honolulu: University of Hawaii Press.

Jessop, Bob (1997). 'A neo-Gramscian Approach to the Regulation of Urban Regimes: Accumulation Strategies, Hegemonic Projects, and Governance'. In M. Lauria ed. *Reconstructing Urban Regime Theory: Regulating Urban Politics in a Global Economy*. Thousand Oaks, CA: Sage Publications, pp. 51–73.

Jessop, Bob (1998). 'The Rise of Governance and the Risks of Failure: The Case of Economic Development'. *International Social Science Journal* 50 (1), pp. 29–45.

Jimenez, E. (1987). 'The Magnitude and Determinants of Home Improvement in Self-Help Housing: Manila's Tondo Project'. *Land Economics* 59, pp. 70–83.

Jomo, K.S. (1998). 'Malaysian Debacle: Whose Fault?'. *Cambridge Journal of Economics* 22 (6), pp. 707–22.

Jones, Alice (1996). 'The Psychology of Sustainability: What Planners Can Learn From Attitude Research'. *Journal of Planning Education and Research* 16, pp. 56–65.

Jones, Martin (1998). 'Restructuring the Local State: Economic Governance or Social Regulation?'. *Political Geography* 17 (8), pp. 959–88.

Jumbala, P. (1987). 'Interest and Pressure Groups'. In S. Xuto, ed. *Government and Politics of Thailand*. Singapore: Oxford University Press, pp. 110–67.

Kalbermatten, J. M., D.S. Julius, C.G. Gunnerson, and D.D. Mara (1982). *Appropriate Sanitation Alternatives: Planning and Design Manual*. Washington, DC: World Bank.

Kemp, Jeremy (1984). 'The Manipulation of Personal Relations: From Kinship to Patron-Clientage'. In Han ten Brummelhuis and Jeremy H. Kemp, eds. *Strategies and Structure in Thai Society*. Amsterdam: Anthropological-Sociological Centre, pp. 55–71.

Keyes, C.F. (1987). *Thailand, Buddhist Kingdom as Modern Nation-State*. Boulder, CO: Westview Press.

Kim, Eun Mee, ed. (1998). *The Four Asian Tigers: Economic Development and the Global Political Economy*. San Diego, CA: Academic Press.

Kim, W.B., M. Douglass and K.C. Ho (1997). *Culture and the City in East Asia*. Oxford: Oxford University Press.

Kingdon, John W. (1984). *Agendas, Alternatives, and Public Policies*. Boston, MA: Little, Brown and Company.

Klausner, William J. (1983). *Reflections on Thai Culture*, 2nd edition. Bangkok: Siam Society.

Klundert A.V.D. and I. Lardinois (1995). 'Community and Private (Formal and Informal) Sector Involvement in Municipal Solid Waste Management in Developing Countries'. Background Paper for the Urban Management Programme (UMP), UNDP Meeting, Ittingen, Switzerland, 10–12 April.

Korten, David C. (1990). *Getting to the 21st Century: Voluntary Action and the Global Agenda*. West Hartford, CT: Kumarian Press.

Kristoff, Nicholas D. (1997). 'Poisoned Lands: Across Asia, a Pollution Disaster Hovers'. *New York Times*, 28 November, pp. A1 and A14.

Krongkaew, Medhi (1990). 'Study Area 4: Urban Finance and Resource Mobilization', Background Report 4–2, National Urban Development Policy Framework, Thailand Development Research Institute Foundation, Bangkok.

Krueger, Jonathan (1999). 'What's Become of Trade in Hazardous Wastes?: The Basel Convention One Decade Later'. *Environment* **41**, pp. 10–18.

Lake, Robert W. (1993). 'Rethinking NIMBY'. *Journal of the American Planning Association* **59**, pp. 87–93.

Laothamatas, Anek (1992). *Business Associations and the New Political Economy of Thailand: from Bureaucratic Polity to Liberal Corporatism*. Boulder, CO: Westview Press; Singapore: Institute of Southeast Asian Studies.

Lapid, D. (1999). 'SWM Privatisation in the Philippines and its Effects on Micro- and Small Enterprises'. Manila: Center for Advanced Philippine Studies.

Laurel, Herman Tiu (2001). 'The Burning Issue' (Opinion). *Manila Times*, 15 January, p. 7.

Lawrence, A.R., D.C. Gooddy, P. Kanathareana, W. Meesilp and V. Ramnarong (2000). 'Groundwater Evolution beneath Hat Yai, a Rapidly Developing City in Thailand'. *Hydrogeology Journal* (September, published online).

Lee, Yok-shiu F. (1998). 'Intermediary Institutions, Community Organizations, and Urban Environmental Management: The Case of Three Bangkok Slums'. *World Development* **26** (6), pp. 993–1011.

Lee, Yok-shiu F. and Alvin Y. So, eds. (1998). *Asia's Environmental Movements: Comparative Perspectives*. Armonk, NY: M.E. Sharpe.

Leinbach, Thomas R. (2000). 'Indonesia in the Financial Crisis'. In Thomas R. Leinbach and Richard Ulack, eds. *Southeast Asia: Diversity and Development*. Upper Saddle River, New Jersey: Prentice Hall, p. 329.

Leinbach, T.R. and R. Ulack, eds (2000). *Southeast Asia: Diversity and Development*. Upper Saddle River, New Jersey: Prentice Hall.

Leitner, Helga (1990). 'Cities in Pursuit of Economic Growth: The Local State as Entrepreneur'. *Political Geography Quarterly* **9** (2), pp. 146–70.

Levy, Daniel C. (1996). *Building the Third Sector: Latin America's Private Research Centers and Nonprofit Development*. Pittsburgh, PA: University of Pittsburgh Press.

Lim, J. (1998). 'The Philippines and the East Asian Economic Turmoil'. In K.S. Jomo, ed. *Tigers in Trouble*. London: Zed Books, pp. 224–43.

Lober, Douglas J. (1997). 'Explaining the Formation of Business-Environmentalist Collaborations: Collaborative Windows and the Paper Task Force'. *Policy Sciences* **30**, pp. 1–24.

Logan, John R. and Harvey Molotch (1987). *Urban Fortunes*. Berkeley, CA: University of California Press.

Luanratana, Watana (1994). Personal Interview. Director, Infrastructure and Environment Planning Division, Department of Policy and Planning, Bangkok Metropolitan Administration, 22 December.

MacLeod, G. (1999). 'Entrepreneurial Spaces, Hegemony, and State Strategy: The Political Shaping of Privatism in Lowland Scotland'. *Environment and Planning A* **31**, pp. 345–75.

MacLeod, Gordon and Mark Goodwin (1999). 'Reconstructing an Urban and Regional Political Economy: On the State, Politics, Scale, and Explanation'. *Political Geography* **18** (6), pp. 697–730.

Magno, Francisco A. (1998). 'Environmental Movements in the Philippines'. In Yok-shiu Lee and Alvin So, eds. *Asia's Environmental Movements: Comparative Perspectives*. Armonk, NY: M.E. Sharpe, pp. 143–78.

Majumdar, T.K. (1995). 'Social Networks, People's Organization and Popular Participation – Process, Mechanisms and Forms in the Squatter Settlements'. *Norsk Geograisk Tidsskrift* **49** (9), pp. 161–76.

Martin, Peter and Helen Ritchie (1999). 'Logics of Participation: Rural Environmental Governance under Neo-liberalism in Australia'. *Environmental Politics* **8** (2), pp. 117–35.

McGee, G. (1967). *The Southeast Asian City: A Social Geography of the Primate Cities of Southeast Asia*. London: Bell; New York: Frederick A. Praeger.

McLeod, R. and Ross Garnaut, eds. (1998). *East Asia in Crisis: From being a miracle to needing one?* London: Routledge.

Meagher, K. (1995). 'Crisis, Informalization and the Urban Informal Sector in Sub-Saharan Africa'. *Development and Change* **26**, pp. 259–84.

Midgley, J., A. Hall, M. Hardiman, and D. Narine (1986). *Community Participation, Social Development and the State*. New York: Methuen.

Minkoff, D. (1997). 'Producing Social Capital: National Social Movements and Civil Society'. *American Behavioral Scientist* **40** (5), pp. 606–19.

Mitchell, K. (1995). 'Flexible Circulation in the Pacific Rim: Capitalism in Cultural Context'. *Economic Geography* **71** (4), pp. 365–83.

Mizuno, Koichi (1976). 'Thai Patterns of Social Organizations: Notes on a Comparative Study'. In Sinichi Ichimura, ed. *Southeast Asia: Nature, Society and Development*. Honolulu: University of Hawaii Press, pp. 79–93.

Mollenkopf, John H. (1983). *The Contested City*. Princeton, NJ: Princeton University Press.

Morgan, K. (1997). 'The Learning Region: Institutions, Innovation and Regional Renewal'. *Regional Studies* **31**, pp. 491–503.

Moser, C.O.N. (1993). 'Community Participation in Urban Projects in the Third World'. *Voices From the City* **2**, pp. 8–9.

Mulder, J.A. Niels (1985). *Everyday Life in Thailand: An Interpretation*, 2nd edition. Bangkok: Editions Duang Kamol.

Mullings, Beverley (1999). 'Sides of the Same Coin? Coping and Resistance among Jamaican Data-Entry Operators'. *Annals of the Association of American Geographers* **89** (2), pp. 290–311.

Mydans, S. (2000) 'Before Manila's Garbage Hill Collapsed: Living off Scavenging (200 Bodies have been Recovered from Village of Squatters)'. *New York Times*, 18 July, p. A6 (N).

Nakata, T. (1987). 'Political Culture: Problems of Development of Democracy'. In S. Xuto, ed. *Government and Politics of Thailand*. Singapore: Oxford University Press, pp. 168–95.

National Economic and Social Development Board (NESDB) (1991). The Seventh National Economic and Social Development Plan. Bangkok.

National Science Foundation (NSF) Workshop on Urban Sustainability (2000). 'Towards a Comprehensive Geographical Perspective on Urban Sustainability'. Final Report of the 1998 National Science Foundation Workshop on Urban Sustainability. New Brunswick, NJ: Center for Urban Policy Research, Rutgers University.

Neher, Clark D. (1999). *Southeast Asia in the New Internal Era*. Boulder, CO: Westview Press.

Ngoc, Mai (2001). 'Motor Boom has Life of its Own'. Vietnam Investment Review: Insight/No. 487 (12–18 February), www.wir-vietnam.com/vir/487int00.html.

Nguyen, D.T. (1997). 'Current situation of air pollution in Ho Chi Minh City, Vietnam'. Proceedings of the Asia-Pacific Conference on Sustainable Energy and Environment Technology, 19–21 June, Singapore, pp. 242–48.

O'Connor, David (1994). 'Managing the Environment with Rapid Industrialization: Lessons from the East Asia Experience'. Paris: Organization for Economic Cooperation and Development.

Omer, M.I.A. (1990). 'Child Health in the Spontaneous Settlements around Khartoum'. *Environment and Urbanization* **2** (2), pp. 65–70.

O'Rourke, D. (2001a). 'Community-driven Regulation: Towards an Improved Model of Environmental Regulation in Vietnam'. In P. Evans, ed. *Livable Cities: The Politics of Urban Livelihood and Sustainability*. Berkeley, CA: University of California Press, pp. 81–115.

O'Rourke, D. (2001b). *Motivating a Conflicted Environmental State. Organization and Environment*. Thousand Oaks, CA: Sage Publications.

Panswad, T. (1987). 'Domestic Wastewaters and Water Pollution Problems in Bangkok and its Vicinity'. Bangkok: Prepared for the National Environment Board.

Pasuk, Phongpaichit and Chris Baker (1995). *Thailand, Economy and Politics*. Kuala Lumpur, Oxford and New York: Oxford University Press.

Pescuma, A. and M. Guaresti (1991). 'Gran Buenos Aires: contaminación y saneamiento'. *Medio Ambiente y Urbanización* **2**, pp. 52–74.

Philippine Center for Investigations (2001). Firm Linked to Estrada got Metro Manila Garbage Contract, www.pcij.org/stories/2001/garage.html (24 January).

Pile, S. and M. Keith, eds (1997). *Geographies of Resistance*. New York: Routledge.

Pistor, Katharina and Philip Wellons (1999). 'The Role of Law and Legal Institutions in Asia's Economic Development, 1960–1995'. New York: published for the Asian Development Bank by Oxford University Press.

Pongpao and Ngamkham (2000). 'Environmental Problems Surface Again'. *Bangkok Post*, 24 February, p. 1.

Ponsapich, Amara (1996). 'Nongovernmental Organizations in Thailand'. In Tadashi Yamamoto, ed. *Emerging Civil Society in the Asia Pacific Community*. Singapore: Institute for Southeast Asian Studies and the Japan Center for International Exchange, pp. 245–70.

Portes, A. and P. Landolt (1996). 'The Downside of Social Capital'. *American Prospect* **26** (May–June), pp. 18–21.

Pratt, James H. (1993). 'Toward Collaborative Evaluation of Community Development NGOs in Thailand'. Vancouver, British Columbia, Canada: Asian Urban Research Network, Centre for Human Settlements, School of Community and Regional Planning, University of British Columbia.

Pulido, Laura (1996). *Environmentalism and Economic Justice*. Tucson, AZ: University of Arizona Press.

Putnam, R.D. (1993). *Making Democracy Work*. Princeton, NJ: Princeton University Press.

Quigley, Kevin (1996). 'Environmental Organizations and Democratic Consolidation in Thailand'. *Crossroads* **9** (2), pp. 1–29.

Raco, Mike and Rob Imrie (2000). 'Governmentality and Rights and Responsibilities in Urban Policy'. *Environment and Planning A* **32** (12), pp. 187–204.

Rhodes, R.A.W. (1997). *Understanding Governance: Policy Networks, Governance, Reflexivity, and Accountability*. Buckingham, PA: Open University Press.

Rigg, Jonathan, ed. (1995). *Counting the Costs: Economic Growth and Environmental Change in Thailand*. Singapore: Institute of Southeast Asian Studies.

Rigg, Jonathan (1997). *Southeast Asia: The Human Landscape of Modernization and Development*. London: Routledge.

Riggs, Fred W. (1966). *Thailand: The Modernization of a Bureaucratic Polity*. Honolulu: East-West Center Press.

Ritchey-Vance, Marion (1991). *The Art of Association: NGOs and Civil Society in Colombia*. Rosslyn, VA: Inter-American Foundation.

Rodan, Garry (1997). *The Prospects for Civil Society in Southeast Asia*. North York, Ontario : University of Toronto-York University Joint Centre for Asia Pacific Studies.

Rodan, Garry, Kevin Hewison and Richard Robison, eds. (1997). *The Political Economy of South-East Asia*. Melbourne: Oxford University Press.

Rose, N. and P. Miller (1992). 'Political Power Beyond the State: Problematics of Government'. *British Journal of Sociology* **43** (2), pp. 173–205.

Ruland, J. (1992). 'Municipal Government and Development in Chiang Mai'. In J. Ruland, ed. *Urban Development in Southeast Asia: Regional Cities and Local Government*. Boulder, CO: Westview Press.

Salamon, Lester M. (1995). *Partners in Public Service: Government-Nonprofit Relations in the Modern Welfare State*. Baltimore, MD: Johns Hopkins University Press.

Samudavanija, C. (1987). 'The Bureaucracy'. In S. Xuto, ed. *Government and Politics of Thailand*. Singapore: Oxford University Press, pp. 75–109.

Scott, J.C. (1985). *Weapons of the Weak: Everyday Forms of Peasant Resistance*. New Haven, CT: Yale University Press.

Seager, Joni (1995). *The New State of the Earth Atlas*. New York: Touchstone.

Setchell, Charles (1992). 'Final Report of the Greater Bangkok Slum Housing Market Study, Volume 1'. Bangkok: Regional Housing and Urban Development Office, United States Agency for International Development.

Skocpol, T. (1985). 'Bringing the State Back In: Strategies of Analysis in Current Research'. In P. B. Evans, D. Rueschemeyer, and T. Skocpol, eds. *Bringing the State Back In*. Cambridge: Cambridge University Press, pp. 3–43.

Smutny, G. and L.M. Takahashi (1999). 'Economic change and environmental conflict in the Western Mountain States of the USA'. *Environment and Planning A* **31** (6), pp. 979–95.

So, Alvin Y. and Yok-shiu F. Lee (1998). 'Environmental Movements in Thailand'. In Yok-shiu Lee and Alvin So, eds. *Asia's Environmental Movements: Comparative Perspectives*. Armonk, NY: M.E. Sharpe, pp. 120–42.

Sriswaskraisorn and Bamroong-ua (1994). 'Bangkokians to Pay for Wastewater'. *Bangkok Post* **49** (364), 30 December, pp. 1–3.

Storper, M. (1997). *The Regional World*. New York: Guilford.

Struyk, R.J, M.L. Hoffman and H.M. Katsura (1990). *The Market for Shelter in Indonesian Cities*. Washington, DC: Urban Institute Press.

Swyngedouw, E.A. (1995). 'The Contradictions of Urban Water Provision: A Study of Guayaquil, Ecuador'. *Third World Planning Review* **17** (4), pp. 387–405.

Takahashi, Lois M. and Gayla Smutny (1998). 'Community Planning for HIV/AIDS Prevention in Orange County, California'. *Journal of the American Planning Association* **64** (4), pp. 441–56.

Takahashi, Lois M. and Rigoberto Rodriguez (2002). 'Access Redefined: Service Pathways of Persons Living with HIV'. *Culture, Health and Sexuality* **4** (1), pp. 67–83.

TDRI (1999). 'Social Impacts of the Asian Economic Crisis in Thailand, Indonesia, Malaysia and the Philippines'. Bangkok: Thailand Development Research Institute.

ten Brummelhuis, Han (1984). 'Abundance and Avoidance: An Interpretation of Thai Individualism'. In Han Ten Brummelhuis and Jeremy H. Kemp, eds. *Strategies and Structure in Thai Society*. Amsterdam: Anthropological-Sociological Centre, pp. 39–54.

Tendler, Judith (1997). *Good Government in the Tropics*. Baltimore, MD: Johns Hopkins University Press.

Towprayoon, S., M. Kozlov and T. Kaeowjaroon (1997). 'Application of Mapping for Assessment of Air Pollution in a Big City'. Proceedings of the Asia-Pacific Conference on Sustainable Energy and Environmental Technology, 19–21 June, Singapore, pp. 249–55.

Trébuil, Guy (1995). 'Pioneer Agriculture, Green Revolution and Environmental Degradation in Thailand'. In Jonathan Rigg, ed. *Counting the Costs: Economic*

Growth and Environmental Change in Thailand. Singapore: Institute of Southeast Asian Studies, pp. 67–89.

Turner, J. (1969). 'Uncontrolled Urban Settlement: Problems and Policies'. In G. Breese, ed. *The City in Newly Developing Countries: Reading on Urbanism and Urbanization*. Englewood Cliffs, NG: Prentice-Hall, pp. 507–34.

Ulack, R. and Gyula Paver (1989). *Atlas of Southeast Asia*. New York: MacMillan.

UNEP (United Nations Environment Programme) (1999). *Global Environment Outlook 2000*. Earthscan Publications, Ltd. London.

UNESCAP (United Nations Economic and Social Commission for Asia and the Pacific) (1995). *State of the Environment in Asia and the Pacific*. Bangkok: United Nations Economic and Social Commission for Asia and the Pacific.

UNESCAP (United Nations Economic and Social Commission for Asia and the Pacific) (1998). *Water Resources Series, No. 79, Guidebook to Water Resources, Use and Management in Asia and the Pacific, Vol II: Water Management*. Bangkok: United Nations Economic and Social Commission for Asia and the Pacific.

UNESCAP (United Nations Economic and Social Commission for Asia and the Pacific) (1999). *Integrating Environmental Considerations into Economic Policy Making: Institutional Issues*. Development Papers, No. 21.

Unkulvasapaul, M. Seidel, H.F. (1991) *Urban Sewage and Wastewater Management in Sector Development. Volume 1: Summary and Main Report*. UNDP/World Bank, Bangkok.

Unger, Danny (1998). *Building Social Capital in Thailand: Fibers, Finance and Infrastructure*. Cambridge: Cambridge University Press.

Vicentian Missionaries (1998). 'The Payatas Environmental Development Programme: Micro-enterprise Promotion and Involvement in Solid Waste Management in Quezon City'. *Environment and Urbanization* **10** (2), pp. 55–68.

Vietnam Daily News (2001). 'Residents Unhappy with Industrial Pollution'. 30 June, p. 3.

Wancharoen, M. (2000). 'Reactor Construction Moving Ahead'. *Bangkok Post,* **55** (56), 25 February, p. 3.

White, A.T. (1981). Community Participation and Education in Community Water Supply and Sanitation Programmes: Methods and Strategies, IRC Technical Paper, No. 12. Rijswijk, the Netherlands: IRC.

White, Gilbert F., David J. Bradley and Anne U. White (1972). *Drawers of Water*. Chicago, IL: University of Chicago Press.

Whittington, D., J. Davis, H. Miarsono and R. Pollard (2000). 'Designing a "Neighborhood Deal" for Urban Sewers: A Case Study of Semarang, Indonesia'. *Journal of Planning Education and Research 19(3), pp. 297–308*.

Whittington, D., D.T. Lauria, K. Choe, J.A. Hughes and V. Swarna (1993). 'Household Sanitation in Kumasi, Ghana: A Description of Current Practices, Attitudes, and Perceptions'. *World Development* **21**, pp. 733–48.

Whittington, D., D.T. Lauria and X. Mu (1991). 'A Study of Water Vending and Willingness to Pay in Onitsha, Nigeria'. *World Development* **19**, pp. 179–98.

WHO and UNEP (World Health Organization and United Nations Environment Program) (1992). *Urban Air Pollution in Megacities of the World*. Cambridge: Blackwell.

Whyte, A.V. (1984). 'Community Participation: Neither Panacea or Myth'. In P. Bourne, ed. *Water and Sanitation: Economic and Sociological Perspectives.* New York: Academic Press, pp. 205–21.

Wolch, Jennifer R. (1989). 'The Shadow State: Transformations in the Voluntary Sector'. In Jennifer Wolch and Michael Dear, eds. *The Power of Geography: How Territory Shapes Social Life.* Boston: Unwin Hyman, pp. 197–221.

Wolch, Jennifer (1990). 'The Shadow State: Government and Voluntary Sector in Transition'. New York: Foundation Center.

Wolch, J. and G. DeVerteuil (1999). 'New Landscapes of Urban Poverty Management'. In Thrift, N. and J. May, eds. *TimeSpace.* London: Routledge.

Woolcock, M. (1998). 'Social Capital and Economic Development: Toward a Theoretical Synthesis and Policy Framework'. *Theory and Society* **27**, pp. 151–208.

World Bank (1990). *World Development Report 1990: Poverty.* New York: Oxford University Press.

World Bank (1998a). *Pollution Prevention and Abatement Handbook.* Washington, DC: World Bank.

World Bank (1998b). *East Asia: The Road to Recovery.* Washington, DC: World Bank.

World Bank (1999). *World Development Indicators.* Washington, DC: World Bank.

Worldwatch Institute (1999). *The World Watch Reader on Global Environmental Issues.* New York: Norton.

WRI (1998). *World Resources 1998–1999.* New York: Oxford University Press.

Wurfel, David and Bruce Burton (1996). *Southeast Asia in the New World Order: The Political Economy of a Dynamic Region.* Houndmills, Basingstoke: Macmillan.

Xuto, S., ed. (1987). *Government and Politics of Thailand.* Singapore: Oxford University Press.

Yacoob, M., B. Braddy and L. Edwards (1992). 'Rethinking sanitation: Adding Behavioral Change to the Project Mix, WASH Technical Report No. 72'. Washington, DC: Water and Sanitation for Health Project.

Yamamoto, Tadashi and Kim Gould Ashizawa, eds (1999). *Corporate-NGO Partnership in Asia Pacific.* Tokyo: Japan Center for International Exchange.

Young, Oran R. and George Demko (1996) 'Improving the Effectiveness of International Environmental Governance Systems'. In Oran R. Young, George J. Demko and Kilaparti Ramakrishna, eds. *Global Environmental Change and International Governance.* Hanover, NH: Dartmouth College, University Press of New England, pp. 215–47.

Name Index

Subject Index

167